MEETING PLACES: SCIENTIFIC CONGRESSES AND URBAN IDENTITY IN VICTORIAN BRITAIN

Science, Technology and Culture, 1700–1945

Series Editors

David M. Knight
University of Durham

and

Trevor Levere
University of Toronto

Science, Technology and Culture, 1700–1945 focuses on the social, cultural, industrial and economic contexts of science and technology from the 'scientific revolution' up to the Second World War. It explores the agricultural and industrial revolutions of the eighteenth century, the coffee-house culture of the Enlightenment, the spread of museums, botanic gardens and expositions in the nineteenth century, to the Franco-Prussian war of 1870, seen as a victory for German science. It also addresses the dependence of society on science and technology in the twentieth century.

Science, Technology and Culture, 1700–1945 addresses issues of the interaction of science, technology and culture in the period from 1700 to 1945, at the same time as including new research within the field of the history of science.

Also in the series

Sir James Dewar, 1842–1923
A Ruthless Chemist
J.S. Rowlinson

Engineer of Revolutionary Russia
Iurii V. Lomonosov (1876–1952) and the Railways
Anthony Heywood

Selling Science in the Age of Newton
Advertising and the Commoditization of Knowledge
Jeffrey R. Wigelsworth

Meeting Places: Scientific Congresses and Urban Identity in Victorian Britain

LOUISE MISKELL
Swansea University, UK

Routledge
Taylor & Francis Group
LONDON AND NEW YORK

First published 2013 by Ashgate Publishing

2 Park Square, Milton Park, Abingdon, Oxon OX14 4RN
711 Third Avenue, New York, NY 10017, USA

Routledge is an imprint of the Taylor & Francis Group, an informa business

First issued in paperback 2016

British Library Cataloguing in Publication Data
Miskell, Louise.
Meeting places : scientific congresses and urban identity in Victorian Britain. – (Science, technology and culture, 1700-1945)
1. British Association for the Advancement of Science–Congresses. 2. National Association for the Promotion of Social Science (Great Britain)–Congresses. 3. Royal Archaeological Institute (Great Britain)–Congresses. 4. Royal Agricultural Society of England–Congresses. 5. Learning and scholarship–Great Britain–History–19th century. 6. Great Britain–Intellectual life–19th century. 7. City promotion–Great Britain–History–19th century.
I. Title II. Series
306.4'1'0941'09034-dc23

Library of Congress Cataloging-in-Publication Data
Miskell, Louise.
Meeting places : scientific congresses and urban identity in Victorian Britain / by Louise Miskell.
p. cm.
Includes bibliographical references and index.
ISBN 978-1-4094-5237-9 (hbk) 1. Associations, institutions, etc.– Great Britain–History–19th century. 2. Urbanization–Great Britain–History–19th century. I. Title.
HS67.M57 2013
506'.041–dc23

2012034183

ISBN 978-1-4094-5237-9 (hbk)
ISBN 978-1-138-26710-7 (pbk)

Contents

List of Figures and Tables

Figures

Tables

Acknowledgements

This book began life in July 2005 at around the time of the announcement that London had successfully bid to host the 2012 Olympics. The scientific meetings I was reading about at the time were perhaps not the Victorian knowledge equivalents of modern Olympic Games, but many of the themes which surfaced that day – competition, cultural capital, the idea that major events could deliver an urban legacy – struck a chord with me as I thought about the urban impact of hosting the major events of the scientific calendar. A period of archival research commenced over the course of the following summer, with the funding provided by the British Academy's Small Research Grants scheme, for which I am extremely grateful. This support enabled me to visit archives at University College London, the Society of Antiquaries, the Bodleian Library, the Museum of English Rural Life in Reading, the Maritime Archives and Library at Liverpool and Manchester Central Library. I am grateful to the staff in all of these locations for their patient assistance with my requests for material.

A number of papers based on the preliminary findings of this research were subsequently presented at research seminars in the Department of History and Welsh History, Aberystwyth, the Centre for Urban History at Leicester and the Department of History at Swansea, and audiences in attendance at each of these provided helpful comments and observations which influenced the direction of my further research and writing. Colleagues and friends at the annual Urban History Group conferences have shown continual interest and encouragement and I would like to thank them, collectively, for their friendly advice and support in recent years. On a daily basis, colleagues in the Department of History and Classics at Swansea University have been generous with their time and expertise. In particular, I am grateful to Adam Mosley for often alerting me to new publications in the history of science, to Martin Johnes for reading early drafts of some of the chapters and to Huw Bowen and David Turner for their helpful suggestions in the preparation of publication proposals. Without two crucial semesters of research leave, in 2009 and 2011, this book would not have been written and thus I am especially grateful to John Spurr and to Chris Williams, of the College of Arts and Humanities at Swansea University, for looking favourably on my applications for sabbatical leave. Outside of my own department, I have profited from conversations about this project with a number of other historians including Neil Evans, Peter Stead and Lucinda Matthews-Jones. I am also indebted to staff at Ashgate, particularly the commissioning editor and the editors of the Science, Technology and Culture, 1700–1945 series, for their interest in and support for this project.

Closer to home, family members have played an important part too. I would like to thank my siblings, in particular my brother Pete and his family who indirectly did me a great favour by moving into a house close to the Museum of English Rural Life in Reading, and for kindly accommodating me during research visits there. My husband, Mike, stoically put up with the clutter of books, papers and files in our 'shared' home office, and undertook far more than his quota of the domestic chores while I spent hours in front of the PC working on the text. In the final few months, his help with preparing illustrations and tables was invaluable and the completion of the book, like much in my life, would have been far more difficult without him. My greatest debt is to my parents, and especially to my father, Anthony Miskell (12 August 1941–23 November 2012), who taught me to love words and to treasure learning. This book is dedicated to his memory.

List of Abbreviations

BAAS	British Association for the Advancement of Science
BRO	Bristol Record Office
CUHB	Cambridge Urban History of Britain
MCL	Manchester Central Library
MERL	Museum of English Rural Life
NAPSS	National Association for the Promotion of Social Science
NML MAL	National Museums Liverpool, Maritime Archives and Library
RAI	Royal Archaeological Institute of Great Britain and Ireland
RASE	Royal Agricultural Society of England
TWAMS	Tyne and Wear Archives and Museums Service
UCL	University College London
WSL	Westcountry Studies Library

Introduction

On Friday 24 August 1838, the editors of the *Newcastle Courant* reflected on the events of the foregoing week in their town. They declared that, 'Newcastle has never before been thronged with so many visitors'. Among their number were such distinguished figures as the Duke and Duchess of Northumberland, the Marquis of Northampton, the members of parliament for the district and a host of eminent scientists including Herschel, Sedgwick and Buckland. By day they had flocked in their hundreds to the Central Exchange, the New Music Hall, the Literary and Philosophical Society, the County Court Rooms, the Surgeon's Hall and the Academy of Arts to hear lectures, debates and reports by the leading scientific figures of the time. By night the Old Circus, the Large Assembly Room and the Green Market, decorated with ornamental lights, fountains, evergreens and flowers, accommodated them for lavish dining, promenading and a ball hosted by the mayoress. In case visitors indulged themselves too much, the entertainment was juxtaposed with a range of more thought-provoking diversions including an exhibition of models at which manufacturers, inventors, designers and engineers showcased specimens of their latest plans and ideas,[1] and promenaders at the Green Market were said to have been fascinated by a 'calculating boy' who demonstrated impressive feats of numeracy in response to questions from assembled guests. At the end of the week, a day was set aside for visits to places of interest in the wider region. Some 200 visitors embarked on a 'geological excursion' to the coast at Tynemouth where they breakfasted in a specially erected marquee before setting out to inspect the coastal geology.[2] Meanwhile another 400 set out in special trains to view the new bridge and railway at Durham Junction, their departure cheered by crowds of spectators and accompanied by the firing of cannons, flag-waving and music from the South Shields band.[3]

The occasion for this heady mixture of education, entertainment, carnival and day-trip, all overlaid with a generous dose of civic pomp, was the eighth annual meeting of the British Association for the Advancement of Science. British towns had been accustomed to hosting scientific visitors of one kind or another long before the Victorian era. Since the early eighteenth century, peripatetic speakers on scientific topics were a common sight in towns, where theatres doubled as lecture halls and low admission prices drew in crowds of eager listeners for whom a talk

---

1 *Newcastle Courant*, 7 September 1838.

2 *Newcastle Courant*, 24 August 1838.

3 *Newcastle Courant*, 31 August 1838.

on the latest scientific ideas by a visiting orator was popular entertainment.[4] By the second half of the eighteenth century, the potential of knowledge to improve, enlighten and enhance status was being harnessed by the resident commercial and professional populations of England's county towns in a proliferation of local literary and philosophical institutions.[5] But all of this provincial activity was not matched in the metropolis where the Royal Society and the Society of Antiquaries, two long-established foci for the intellectual elite, were in a state of stagnation, rendered impotent by restricted membership and inactivity.[6] The foundation in 1831 of the British Association for the Advancement of Science, followed, in the early years of Victoria's reign, by other national organisations for the promotion of agriculture, medicine, social science, archaeology and a myriad of other branches of knowledge, came about in response to this lacunae at the centre. The BAAS, in particular, was founded in the midst of debate about a decline of science in England. The need for new stimulus and direction was an issue of national reputation for scientists like Charles Babbage and John Herschel.[7] Recognising the vitality of provincial knowledge culture, these new national associations sought to harness and direct it. The main strategy they adopted for achieving this was the annual meeting: a week-long congress during which the association and its delegates descended on a nominated town or city and held a programme of lectures, discussions, exhibitions and social functions in the largest available halls and reception rooms. The aim was to 'Let Philosophy at length come forth and show herself in public; let her hold her court in different parts of her dominions'.[8] By the mid-Victorian years this strategy, set out in York in 1831 by William Harcourt,[9] for the newly inaugurated British Association, had become had become a familiar feature of the summer months, and Britain's provincial towns and cities, where knowledge culture was thriving, became the platform for advancing knowledge on the national stage. The annual British Association meetings earned the nickname 'parliaments of science', but the tag was also applied more generally to the yearly

---

4 G.S. Kitteringham, 'Studies in the Popularisation of Science in England, 1800–1830', unpublished Ph.D. thesis (University of Kent, 1981), pp. 280–281.

5 Paul Elliott, 'Towards a geography of English scientific culture: provincial identity and literary and philosophical culture in the English County town, 1750–1850', *Urban History*, 32/3 (2005): 391–412.

6 Howard Mumford Jones and I. Bernard Cohen (eds), *Science Before Darwin. A Nineteenth-Century Anthology* (London, 1963), p. 297; Rosemary Sweet, *Antiquaries. The Discovery of the Past in Eighteenth-Century Britain* (London, 2004), pp. 109–10.

7 Colin A. Russell, *Science and Social Change, 1700–1900* (London, 1983), pp. 174–92.

8 Excerpt from speech by William Harcourt, proposing the formation of the British Association for the Advancement of Science, quoted in Osbert J.R. Howarth, *The British Association for the Advancement of Science: a Retrospect, 1831–1931* (London, 1931), p. 21.

9 Jack Morrell, 'Harcourt, William Venables Vernon (1789–1871)', *Oxford Dictionary of National Biography* (Oxford, online edition, 2004) accessed 6 July 2009.

gatherings of other big knowledge associations which adopted the model of the town-based annual meeting for the promotion, communication and advancement of the latest ideas in their discipline. An editorial published in *The Times* in 1878 anticipated the forthcoming conference season:

> The peripatetic parliaments of science will be on us immediately, the first of them, the British Medical Association, meeting next week at Bath, to be followed in rapid succession by the British Association [for the Advancement of Science], the Social and Sanitary Science, the Archaeological and various other special associations ... The corresponding associations of France and America, both largely modelled after the British prototype, meet a week after in Paris and St Louis respectively.[10]

This is a study of the parliaments of science phenomenon in Victorian Britain. Its aim is to investigate these events for what they reveal about the towns and cities which hosted them and to explore the ways in which they could influence the evolving identity and status of nineteenth-century towns. They belonged to an era when the spread of ideas to the wider public arguably took precedence over communication with intellectual peers. This was the case not just in Britain. In early nineteenth-century America, cities provided a vital context for intellectual life, especially in the era before the widespread specialisation and institutionalisation of learning in universities, as scientists and other intellectuals engaged in public knowledge forums as well as in their own professional knowledge networks.[11] In Britain, knowledge cultivation was a key urban function, particularly in the period before the development of the provincial university sector, when the populous towns and cities were where the audiences for new scientific theories were most numerous and where novel intellectual ideas could be most effectively communicated to the public. Annual meetings were by no means the only mechanisms for achieving this. Touring speakers, classes and lectures held at town-based institutions like mechanics' institutes and literary and philosophical societies, and pages of newspapers and more specialist quarterly and monthly journals, were also important vehicles for the dissemination of knowledge.[12] But it was the gathering of otherwise scattered enthusiasts and practitioners in mass associations which was, according to one historian of knowledge in nineteenth-century Britain, 'the characteristic institutional form of the mid-Victorian period',[13] and it was the annual meetings of these associations which were most visible to the

---

10 *The Times*, 3 August 1878.

11 See Thomas Bender, *Intellect and Public Life. Essays on the Social History of Academic Intellectuals in the United States* (Baltimore, 1993), pp. 20–27.

12 Martin Daunton, 'Introduction', in Martin Daunton (ed.), *The Organisation of Knowledge in Victorian Britain* (Oxford, 2005), pp. 4–5.

13 Lawrence Goldman, *Science, Reform and Politics in Victorian Britain. The Social Science Association, 1857–1886* (Cambridge, 2002), p. 370.

public at large. This was not an exclusively British phenomenon. An editorial in *The Times* in 1860 noted the ubiquity of the practice:

> The scientific or artistic congress is an institution which has become most popular in Europe during the last few years. Every department of human thought now has its annual or occasional gathering, at which the votaries, professionals and amateurs, report, progress, plan new undertakings, [and] give their ideas to the world.[14]

In a British context, however, the parliaments of science can be regarded almost as an emblem of the age, made possible by the great railway expansion, which commenced in the 1830s, and given impetus by the Victorian thirst for knowledge, especially science. The meetings flourished particularly during the mid-Victorian 'equipoise' of the 1850s and 1860s when threats of urban disorder had receded, and the large provincial centres of population presented a more inviting environment for the communication of the intellectual ideas of the day. But the organisation and operation of these great scientific meetings also demonstrated many of the characteristic tensions of an age when opposing interests were finely counter-balanced:[15] the relationship between centre and provinces, the eclipse of the county and market town by the industrial metropolis, the pressure in intellectual life between amateur and professional, and specialist versus recreational science. To examine the parliaments of science, then, is to engage with issues central to understanding Victorian society.

By the last quarter of the nineteenth century, the range of different groups and associations using town-based annual meetings as their favoured means of communicating their ideas to their members and to the public was large and diverse. Richard Rodger's study of civic hospitality expenses in late nineteenth-century Edinburgh lists, among the associations visiting the city from 1879–1913, a host of professional bodies including the British Medical Association, the British Dental Association, the Institute of Journalists, the Institute of Mechanical Engineers and the Royal Institute of British Architects. Friendly society gatherings were also hosted, including the Independent Order of Good Templars and the Independent Order of Rechabites, as were a clutch of charitable and religious organisations, including the British Women's Temperance Association, the Pan

14 *The Times*, 26 September 1860.

15 The 'Age of Equipoise' was 'a generation in which the old and the new, the elements of growth, survival and decay, achieved a balance which most contemporaries regarded as satisfactory'. William Laurence Burn, *The Age of Equipoise. A Study of the Mid-Victorian Generation* (London, 2nd edition, 1968), p. 17. But the degree of tension and flux in these decades should not be understated and is central to Burn's interpretation. See Martin Hewitt, 'Prologue', in Martin Hewitt (ed.), *An Age of Equipoise? Re-assessing Mid-Victorian Britain* (Aldershot, 2000), pp. 1–58.

Congregational Church and the World Women's Christian Temperance Union.[16] In addition, numerous county and regional organisations for the promotion of different branches of knowledge, from the Devonshire Association for the Advancement of Science, to the Bath and West of England Society and the Cambrian Archaeological Association, also used rotating annual meetings as a means of attracting new members and extending their intellectual influence as widely as possible in their specific geographical region. The meeting patterns of any one of these organisations might be examined profitably for evidence of the role of towns as congressional venues in the nineteenth century but, for the purposes of this study, the focus is on the meetings of four organisations which might be regarded as forming a top tier of knowledge associations active in the first half of Victoria's reign. The four, the British Association for the Advancement of Science (BAAS), the Royal Agricultural Society of England (RASE), the Royal Archaeological Institute (RAI) and the National Association for the Promotion of Social Science (NAPSS), shared a number of characteristics. First, they purported to be national in their reach and attempted to extend their activities and membership country-wide. Second, they were prominent in the second and third quarters of the nineteenth century, when the potential of the expanding railway network to facilitate mass gatherings was first being realised, and the heyday of their activities overlapped in the 1850s, 1860s and 1870s. Third, unlike the many professional and specialist associations which flourished in the last quarter of the nineteenth century,[17] these bodies were broad-based in terms of their intellectual interests and their membership was not confined to specific professional, religious or other special interest groups. Finally, the fact that they each adopted the principle of the peripatetic annual meeting, pioneered by the BAAS and consciously followed by the other three associations, gave them a shared operational remit when it came to the dissemination of the latest ideas relating to their particular branch of knowledge.

Authors of histories of these four associations have explored aspects of the organisation and business of their annual congresses. Morrell and Thackray's research on the early years of the BAAS traced the development of its annual meetings from the early dominance of academic and metropolitan centres, to rising commercial cities and, occasionally, provincial outposts.[18] Likewise, the

---

16 Richard Rodger, 'The "common good" and civic promotion: Edinburgh 1860–1914', in Robert Colls and Richard Rodger (eds), *Cities of Ideas. Civil Society and Urban Governance in Britain, 1800–2000* (Aldershot, 2004), p. 148.

17 Professional associations were not unknown in the 1830s and 1840s. For one example see Ernest M. Little, *History of the British Medical Association, 1832–1932* (London, 1932). In general, a proliferation of more narrowly defined, disciplinary associations was a feature of the last quarter of the nineteenth century, Bender, *Intellect and Public Life*, p. 43.

18 Jack Morrell and Arnold Thackray, *Gentlemen of Science. Early Years of the British Association for the Advancement of Science* (Oxford, 1981), pp. 78–108, 166–89.

authors of other association histories have examined the location and function of the annual meeting and its success in spreading the influence of the parent body. The early 'country meetings' of the Royal Agricultural Society, for example, were conceived as 'storm centres' of innovation, where the latest agricultural improvement techniques would be communicated widely to the provincial farming community.[19] The social scientists also took their messages of sanitary health and educational improvement around the country in an attempt to press for legislative reform.[20] The archaeologists, meanwhile, held their annual meetings predominantly in the county towns of England where they tried to 'impress and involve people of power and influence in the locality'.[21] But while annual meetings have been viewed as part of the histories of their parent associations, interest in them beyond this sphere has been limited. Considering the familiar and regular place they came to occupy in the summer and early autumn calendar of the period, parliaments of science have attracted surprisingly little attention from historians of the mid-Victorian era. For Asa Briggs, interest in science and other branches of knowledge was an essential element of 'Victorianism' but one which was measured by the popularity of scientific publications and the trend for collecting and cataloguing specimens, rather than attendance at annual congresses.[22] In G.M. Young's survey of the period, the annual scientific meetings are noticed only as events which attracted sceptical comment in the press,[23] while in other well-known surveys of the Victorian years they receive little if any mention.[24] Another historian of mid-Victorian Britain, W.L. Burn, was said to have been, 'more concerned with enduring institutions and ideas, rather than temporary forms of entertainment or "feasts"'.[25] It was a comment made to explain the absence of attention in Burn's work to the Great Exhibition of 1851, but equally it may help explain why the great knowledge festivals of the day feature little in the classic histories of the period.

Perhaps more surprising is the fact that historians of science have not more readily investigated the parliaments of science phenomenon. The British

---

19 Nicholas Goddard, *Harvests of Change. The Royal Agricultural Society of England, 1838–1988* (London, 1988), p. 31.

20 Goldman, *Science, Reform and Politics*, pp. 91–102.

21 Linda Ebbatson, 'Conditions of Emergence and Existence of Archaeology in the Nineteenth Century: the Royal Archaeological Institute, 1843–1914', unpublished Ph.D. thesis, (University of Durham, 1999), p. 14.

22 Asa Briggs, *The Age of Improvement 1783–1867* (Harlow, 7th edition, 1988), pp .480–483.

23 E.E. Kellett, 'The Press', in G.M. Young (ed.), *Early Victorian England, 1830–1865. Volume II* (London, 3rd edition, 1963), p. 56.

24 See, for example, George S.R. Kitson Clark, *The Making of Victorian England* (London, 1962); Geoffrey Best, *Mid-Victorian Britain, 1851–75* (1971).

25 Peter H. Hoffenberg, 'Equipoise and its discontents: voices of dissent during the international exhibitions', in Hewitt (ed.), *An Age of Equipoise?*, pp. 40–41.

Association's events have received some scholarly attention. Philip Lowe argued that their annual meetings had 'a major continuing impact on provincial science culture and vice versa',[26] while other, more critical studies have pointed to the limitations of the meetings rather than the benefits they brought to science.[27] More recently, new research has yielded fresh insights into the delegate experience at British Association meetings and the local slant that different towns and cities could bring to the 'production and reception of science'.[28] But there have been no attempts to look at annual meetings more broadly, as a subject in their own right. Instead, studies of some of the more influential local scientific societies which sprang up in the provinces, such as the Newcastle Literary and Philosophical Society founded in 1793, and the slightly younger Yorkshire Philosophical Society,[29] have been the preferred method of investigating Britain's burgeoning knowledge culture in the Georgian and Victorian years which amounted, in the words of one scholar, to a second 'scientific revolution'.[30] Provincial scientific establishments like museums, mechanics institutes or literary and philosophical societies, were the most solid, permanent and visible symbols of learning and the growing institutionalisation of knowledge in Britain's towns and cities. Annual parliaments of science, by contrast, were temporary events, visited on towns from outside rather than nurtured from within. This may partly explain why they have escaped scrutiny, in favour of more tangible manifestations of science culture which could be understood as products of the commercial and industrial character of urban communities and their local elites.[31]

[26] Philip Lowe, 'The British Association and the Provincial Public', in Roy MacLeod and Peter Collins (eds), *The Parliament of Science. The British Association for the Advancement of Science, 1831–1981* (Northwood, 1981), p. 119.

[27] A.D. Orange, 'The Idols of Theatre: the British Association and its Early Critics', *Annals of Science*, 32/3 (1975): 278–94.

[28] Rebekah Higgitt and Charles W.J. Withers, 'Science and Sociability. Women and Audience at the British Association for the Advancement of Science, 1831–1901', *Isis*, 99/1 (2008): 1–27; Charles Withers, Rebekah Higgitt and Diarmid Finnegan, 'Historical Geographies of Provincial Science: themes in the setting and reception of the British Association for the Advancement of Science in Britain and Ireland, 1831–1939', *British Journal of the History of Science*, 41/3 (2008): 385–415.

[29] Derek Orange, 'Rational Dissent and Provincial Science: William Turner and the Newcastle Literary and Philosophical Society', in Ian Inkster and Jack Morrell (eds), *Metropolis and Province. Science in British Culture, 1780–1850* (London, 1983), pp. 205–25. Derek Orange, 'Science in early nineteenth-century York: the Yorkshire Philosophical Society and the British Association', in Charles Feinstein (ed.), *York, 1831–1981. 150 Years of Scientific Endeavour and Social Change* (York, 1981), pp. 1–29.

[30] Arnold Thackray, 'Natural Knowledge in Cultural Context: The Manchester Model', *American Historical Review*, 79/3 (1974): 674.

[31] For example, Inkster and Morrell (eds), *Metropolis and Province*, introduction and various chapters.

But the influence of place on the production of knowledge has become a more prominent theme in recent academic scholarship. Literature published in a range of different historical fields reveals the extent to which the exploration of space and place has become a tool for understanding past social and cultural phenomena.[32] The history of science has been no exception. It has long been possible, by comparing studies of local scientific institutions, to see how towns and cities produced different forms of scientific culture. The learned institutions which evolved in Manchester, for example, where industrial growth in the late eighteenth and early nineteenth centuries stimulated scientific enquiry,[33] were very different to those of Norwich with its range of gentlemanly scientific institutions.[34] As part of the 'spatial turn' in science and technology studies, a variety of specific venues have been scrutinised for their role in shaping scientific knowledge, from laboratories, to botanical gardens and museums.[35] Geographical regions, too, have been examined for their tendency to foster particular kinds of scientific investigation according to the flora and fauna, the topography and geology predominating there.[36] Surprisingly, given this new level of interest in the spatial turn, the town is one unit of place which has yet to receive very much investigation as a location for science. Yet, as the opening passage of this book reveals, annual meetings like the British Association's in Newcastle-upon-Tyne in 1838 were occasions when the town itself became a venue for science, propelled onto the national stage. They were events rooted in their urban locations.

Some important observations have been made about the role of towns and cities in shaping scientific activity in the nineteenth and twentieth centuries. In particular, an edition of essays published in 2003 on *Science and the City* contained a number of case studies of scientific investigation in different urban locations throughout Europe, Asia and America. The authors made the case for 'an urban history of science', by demonstrating the active role of cities in shaping scientific knowledge and practice.[37] In this work, in common with other place-focused

32 Ralph Kingston, 'Mind Over Matter? History and the Spatial Turn', *Cultural and Social History*, 17/1 (2010): 111–21.

33 F. Fairbrother, J.B. Birks, W. Mays and P.G. Morgan, 'The History of Science in Manchester', in Charles Frederick Carter (eds), *Manchester and its Region* (Manchester, 1962), pp. 187–97.

34 Trevor Fawcett, 'Popular Science in Eighteenth-Century Norwich', *History Today*, 22 (1972): 590–595.

35 David N. Livingstone, *Putting Science in its Place. Geographies of Scientific Knowledge* (Chicago, 2003); Diarmid A. Finnegan, 'The Spatial Turn: Geographical Approaches in the History of Science', *Journal of the History of Biology*, 41 (2008): 369–88.

36 Simon Naylor, *Regionalizing Science: Placing Knowledges in Victorian England* (London, 2010).

37 Sven Dierig, Jens Lachmund and J. Andrew Mendelsohn (eds), *Science and the City. Osiris*, 18 (2003): pp. 1–19.

science and technology studies, the emphasis is on how place influenced the type of knowledge produced, and the starting point for the analysis is the science rather than the town. In this book, the starting point is reversed: the towns and cities in which annual congresses of the country's leading knowledge associations were staged form the subject of analysis, and it is the lasting influence of this scientific event on the status and identity of the place which is examined. This approach, of examining urban life through the lens of scientific activity, rather than vice-versa, has been used profitably in an American context where the intellectual life of the city has been explored as an essential part of the urban experience. Research on New York in particular, has revealed how the fostering of cultural and educational life became important after the loss of its role as a temporary capital and was used as a means of countering its more prominent commercial image.[38] One of the aims of this book is to show that parliaments of science were used in a similar way by towns and cities in Victorian Britain, to generate cultural capital. Associations with a remit to promote knowledge nation-wide were, potentially, a more powerful and wide-reaching tool of cultural enhancement than provincial scientific institutions or local societies.

National scientific meetings, as well as providing towns with an opportunity to flaunt their credentials as knowledge venues, also reveal much about how urban facilities in the Victorian provinces were viewed and compared by contemporaries. The annual debates over the location and staging of parliaments of science illuminate a number of themes of major importance to urban historians of the period which form the core of this book. It was in the public display of cultural and civic assets that inter-town rivalries were most commonly expressed in urban Britain. In studies of how this operated in practice, historians have drawn attention to the building of ever more lavish and expensive town halls, libraries, parks and museums.[39] In the middle decades of the nineteenth century in particular, these were viewed as the primary hallmarks of urban status in provincial Britain, more so, for instance, than the acquisition of city status, which was not a major goal of ambitious towns until the late nineteenth century.[40] But, especially from the mid-century onwards, in the wake of the 1851 Great Exhibition, the 'fashion' for staging of spectacular, city-based cultural festivals, from sports events, to trade fairs, music competitions and scientific congresses, was also an important status tool used by ambitious towns.[41] Scholarly interest in the impact of grand, international festivals on their host venues has produced a number of comparative studies focusing on the way cities harnessed the power of sporting

---

38 Thomas Bender, *New York Intellect. A History of Intellectual Life in New York City, From 1750 to the Beginning of Our Own Times* (New York, 1987).

39 See for example, Jon Stobart, 'Identity, Competition and Place Promotion in the Five Towns', *Urban History*, 30/2 (2003): 163–82.

40 John V. Beckett, *City Status in the British Isles, 1830–2002* (Aldershot, 2005), p. 19.

41 John R. Gold and Margaret M. Gold, 'Culture and the City', *History in Focus*, 13 (2008), p. 4.

and cultural mega-events in the late nineteenth and twentieth centuries.[42] The very act of bidding to host such an event was a promotional undertaking, in which a town's cultural credentials were set out to be judged and compared with those of competing venues.[43] The competition to host prestigious parliaments of science in mid-Victorian Britain can be viewed as an earlier, more modest incarnation of this phenomenon, but one which provided a similar impetus to town rivalries. The major annual scientific meetings gave provincial towns an opportunity to showcase the range and quality of their local cultural life on a national stage. The sight of the country's leading scientific figures and intellectuals flocking to read lectures and hear debates in their local corn exchange or mechanics' institute reading room, and mixing with the members of their own scientific institutions, was one to which many towns aspired. Consequently, the process of finding towns and cities to host meetings of the major associations became a competitive one. The BAAS and the other associations under scrutiny considered the facilities of the town as an overall place 'package' and made judgements accordingly on whether particular towns or cities constituted the right location for their meetings. Their annual selection contests, conducted with varying levels of rigour and formality, pitted towns and cities against one another as association officials weighed up the relative merits of county towns, ports and industrial metropolises as places in which the great intellectual ideas of the day could be discussed and disseminated.

For the towns and cities which were successful in these competitions, the responsibility for the planning and running of the meeting was onerous. Local organisers shouldered most of the burden of preparation but this gave them scope to tailor the event to suit their own ends. It will be shown in this study that the annual meeting became an important occasion for civic celebration as town leaders strove to gain maximum kudos and visibility. The wearing of official robes, the displaying of corporation symbols, the reading of formal addresses and other civic tools were all utilised by town leaders during meeting weeks to communicate the hospitality and unity of the town in welcoming the visiting association. Other kinds of public ceremonial, celebration and festivity, accompanying events such as the opening of new town halls, the unveiling of statues and the welcoming of royal visitors, have been identified as part of a burgeoning public culture in Britain's provincial cities. Such activities were driven by ambitious urban leaders, eager to proclaim the status of their towns and to stay a step ahead of neighbouring urban centres, vying for recognition on a regional and national stage.[44] Here it is argued that scientific meetings were utilised in a similar way and should be incorporated into this picture

---

[42] See for example, John R. Gold and Margaret M. Gold, *Cities of Culture. Staging International Festivals and the Urban Agenda, 1851–2000* (Aldershot, 2005).

[43] John R. Gold and Stephen V. Ward (eds), *Place Promotion. The Use of Publicity and Marketing to Sell Towns and Regions* (Chichester, 1994), p .2.

[44] See Simon Gunn, *The Public Culture of the Victorian Middle Class. Ritual and Authority in the English Industrial City, 1840–1914* (Manchester, 2nd edition, 2007).

of the emerging provincial 'public culture' which absorbed so much of the energies of urban leaders in the Victorian era.

By examining the public festivities and ceremonial trappings which increasingly accompanied the staging of annual meetings in Victorian towns, this book also offers a different perspective on the question of participation in these flagship scientific events. Historians of science tend to view scientific audiences as consisting of the lecture attendees, book purchasers, patrons or promoters of scientists and their work.[45] By focusing less exclusively on the scientific business and more on the congresses as urban events, it is possible to adopt a broader view of the 'audience' and their levels of involvement. Far from being confined to those who gained access to meeting rooms and lecture theatres, the evidence presented suggests that a much wider cross-section of the community engaged with the spectacle and sense of occasion that accompanied annual meetings. The public was as likely to be excited by the draw of celebrity speakers and the arrival of large numbers of visitors in the town as they were by the news of the latest scientific ideas and theories. Indeed the behaviour of people both inside and outside the meeting rooms often defied the expectations of organisers. The conduct of women, in particular, came under close scrutiny at both the social and scientific elements of the meetings where their behaviour did not always conform to the expectations of male observers.

Ultimately, it was the knowledge that the whole town was on show which gave urban leaders a sense of the significance of these occasions. The performance of the town as a cultural centre capable of hosting a major learned congress could convey a powerful message about urban status to neighbouring towns and cities in the wider region. A varied literature on 'place promotion' has, in recent years, suggested that the creation of new identities was an essential part of the way in which towns and cities promoted themselves on a neighbourhood, regional or national scale.[46] The idea of the town as a centre for science or as another kind of knowledge venue was one such identity and some towns successfully projected this image of themselves in the wake of hosting a scientific congress. Guidebook literature showed towns, where meetings of the BAAS and like organisations had been held, trading on their new-found status as congressional centres long after the event. Scientific meetings also left a physical legacy in the shape of new halls and public landmarks, built and inaugurated to accommodate the meetings, but utilised thereafter for the staging of other cultural events. There were institutional legacies too, with the foundation of local branches of the national associations to continue the work begun at an annual meeting and, in the longer term,

---

[45] See for example, Jan V. Golinksi, 'Utility and Audience in Eighteenth-Century Chemistry: Case Studies of William Cullen and Joseph Priestly', *The British Journal for the History of Science*, 21/1 (1988): 1–31.

[46] Michael Barke and Ken Harrop, 'Selling the industrial town: identity, image and illusion', in Gold and Ward (eds), *Place Promotion*, p. 93.

developments leading towards the establishment of technical colleges and civic universities.

The primary source material available to support a study of this type is rich and varied. In the case of the BAAS and the RASE, extensive institutional archives exist which allow the historian to discover how the towns and cities of Victorian Britain were regarded as potential host venues for annual meetings.[47] Letters and minute books from these collections, recording the views of council members as they debated the relative merits of towns competing for the opportunity to host annual meetings, have been extensively mined for Chapters 1 and 2 of this book. For the RAI and the NAPSS it has been possible to tease out similar kinds of evidence from the surviving letters and personal collections of key figures in the two associations, in particular Lord Brougham, the first president of the NAPSS, and Albert Way of the RAI.[48] Different types of evidence are available for the scrutiny of the way meetings were planned and run by their receiving towns. The deliberations of local organising committees are occasionally preserved in family papers, such as those of John Towne Danson, the Liverpool merchant, who was involved in preparations to host the meeting of the NAPSS in Liverpool in 1858,[49] or in minute books such as that of the Bristol organising committee which prepared for the British Association meeting in their city in 1836.[50] Printed pamphlets and copies of programmes for various meetings often survive. The BAAS deposit at the Bodelian Library, Oxford, for example, includes files of printed material relating to annual meetings since 1831.[51] In some cases, printed ephemera including admission tickets, programmes, maps and menus have been preserved in scrapbooks, also containing newspaper cuttings relating to particular meetings.[52]

Reports printed in provincial newspapers provide the largest body of evidence relating to the meetings, including accounts of the lectures, exhibitions, excursions and dinners which host towns laid on. As with all forms of historical evidence, use of this material is fraught with problems as well as possibilities. Reports were often infused with the political prejudices of editors and their correspondents and

---

47 The archive of the British Association for the Advancement of Science is held in the Bodleian Library Oxford [Dep. BAAS]; that of the Royal Agricultural Society of England is housed at the Museum of English Rural Life [MERL] in Reading.

48 The letters of Lord Brougham form part of the special collections housed at the University College London archives. Way's papers are lodged with the Society of Antiquaries in London.

49 See for example, National Museums Liverpool, Maritime Archives and Library [NML MAL], D/D/III/2/62, letters from George W. Hastings to John T. Danson.

50 Bristol Record Office [BRO], 32079(39) Minute Book of the Reception Committee for the British Association visit.

51 Bodelian, Dep. BAAS 146–55, Printed material for annual meetings, 1831–59.

52 See for example, Manchester Central Library [MCL], BR F306 S3, Social Science Association, Manchester Meeting 1866.

laced with local swagger. But the availability, since 2008, of many nineteenth-century newspapers in digitised form via online databases means that this material is searchable and accessible to a much greater degree than was possible for previous generations of researchers.[53] Use of these digitised daily and weekly papers, supplemented with material from non-digitised titles, forms an invaluable source of evidence on the meetings and constitutes an important part of the material deployed in all of the following chapters. The illustrated press of the mid-Victorian period also provides a valuable source of images of annual meetings. The launch of the *Illustrated London News* in 1842 and its use of images produced by wood engraving brought the sights and scenes of the major scientific congresses of the day to a wide audience. But like the written newspaper reports of the era, these cannot be taken at face value. Produced first and foremost as collectable fine art prints rather than accurate depictions of the events of the day, they were not the work of single, eye-witness artists, but rather of 'teams of engravers, each working on a small area of the total print from sketches that were necessarily approximate'.[54] Many of the examples deployed in this volume emphasised the scale and visual spectacle of the meetings, but must necessarily be treated as an interpretation, rather than a faithful representation of the events.

For evidence of peoples' experience of attending annual meetings and a view of their worth and relevance, newspapers also provide some useful material in the form of letters written to editors by participants and other interested individuals in host towns. More fulsome accounts recorded in diaries, correspondence and reports of people who attended are also available, and range from formal accounts of sectional proceedings, such as that written by Bristol shipping agent Mark Whitwill, who went to the Glasgow Social Science congress in 1874,[55] to the humorous pamphlet, written and illustrated by 'a cheerful pterodactyl', on the 1898 British Association visit to Bristol.[56] This kind of evidence, although patchy and possibly unrepresentative, provides a valuable glimpse of the delegate perspective, and can yield useful insights into the practical effectiveness of the venues used and the interest sparked in the excursions and exhibitions which gave

[53] The British Library's nineteenth-century newspaper database allows the text of 48 provincial daily and weekly newspapers to be searched online for the 1800–1913 period. See *British Library Nineteenth Century Newspapers* at www.galegroup.com. The transformative effect of this on historical research is discussed in Adrian Bingham, 'The Digitization of newspaper archives: opportunities and challenges for historians', *Twentieth Century British History*, 21/2 (2010): 225–31.

[54] See 'Introduction', and Malcolm Chase, '"the Original to the Life": portraiture and the Northern Star', in Laurel Brake and Marysa Demoor (eds), *The Lure of Illustration in the Nineteenth Century* (Basingstoke, 2009), pp.3–4; p.85.

[55] BRO 38605/X/5, Mark Whitwill's Report of the Social Science Congress held at Glasgow on 30 September 1874.

[56] See BRO 29427 (1–6) Booklets and printed materials relating to Bristol meetings of the BAAS in 1898 and 1930.

the peripatetic meetings their distinctive local colour. For analysis of the impact of the meetings on urban identity, guidebooks and handbooks produced at the time of the meetings and afterwards are an important source.[57] It was through printed literature such as this that Victorian towns and cities promoted themselves, and the boost given by the recent visit of a national association is sometimes evident in the pages of such publications. The effect of the meeting on local scientific culture is also considered by tracing the ongoing vitality of town-based scientific societies and, in some cases, the establishment of new knowledge institutions and of more formal educational institutions in the form of colleges and, eventually, civic universities. Less frequently utilised throughout the book are the publications of the associations themselves. The voluminous journals, transactions and reports produced by each of the associations featured in this study provide a body of evidence more representative of the work of the associations and their members than of the towns which hosted their annual congresses. Where this material is deployed in this book, it is for evidence of the extent to which local topics and speakers featured in the scientific business of the meetings.

In order to pursue the key issues identified in this introduction, the book is structured thematically. It begins with a survey of the four featured associations and a comparison of the priorities and practices they employed in deciding on their meeting locations. The different attractions of county towns, ports, resorts and industrial metropolises are explored from the perspective of the four organisations. The following three chapters of the book track the life-cycle of the annual meeting. Beginning in Chapter 2 with the process of bidding to stage an event, which potentially involved intense inter-town competition, there follows an analysis of the organising and running of the meeting in Chapter 3, during which the host venue could deploy all of the civic paraphernalia at its disposal. The experience of attending of these events, when the behaviour and attitudes of different sections of the local populace came under scrutiny, is examined in Chapter 4, and Chapter 5 looks at the legacy annual meetings left behind in host towns and how, in some places, the experience was a catalyst to new initiatives and a strengthened sense of urban identity. The study period adopted reflects the life-span of the associations under scrutiny. 1831, the foundation year of the BAAS, marks the commencement of the analysis and, although the other three associations did not come into being until 1838, 1843 and 1856, there was then a period of almost three decades when all four operated concurrently. The study ends in 1884, the year in which the BAAS took its annual congress overseas for the first time and the NAPSS held its last parliament of science. Although the other associations continued to operate well into the twentieth century, there was a trend, in the mid-1880s, towards more clearly delineated academic disciplines in this period and it therefore forms a natural end-point for the study.

---

[57] For example, Samuel C. Gamwell, *The Official Guide and Handbook to Swansea and its District* (Swansea, 1880).

The five decades covered in this book, however, were ones of significant urban re-alignment in Britain. Demographic change and developments in transport and communications were all making their effects felt on the rank, size, status and functions of towns throughout the country. Long-established county and regional centres were becoming eclipsed by rising industrial heavyweights. Railway lines were bringing new-found prominence to some towns while consigning those that were by-passed to marginality and decline. Urban historians have done much to identify the winners and losers in these processes and the widening gap between the biggest regional centres like Manchester, Birmingham and Liverpool and the more traditional county towns like Canterbury, York and Chester.[58] As new commercial and industrial hubs emerged, the process of re-defining county and regional centres began. Battles over the location of assize courts, the size of town halls and the routing of railway lines as well as the development of civic ceremonial and pageantry, can all be understood in the context of this re-drawing of the urban map of Britain. Ambitious urban leaders seized on opportunities presented by developments such as these to assert the importance of their town over its rival urban centres. Annual scientific meetings, although they have not previously been considered in this context, formed an integral part of this fluctuating picture. The parliaments of science of the early and mid-Victorian era are considered here as indicators of the ambition and status of towns in a period when the hierarchies and functions of Britain's provincial urban centres were undergoing change. From the point of view of a host town, the staging of an annual meeting for a major learned association was a means by which a statement could be made about the importance of a place in the context of its wider county and region. From the point of view of the leading learned associations of Victorian Britain, certain provincial towns and cities, with their concentrated populations, their lecture halls and reception rooms, and their ambitious and status-conscious civic leaders, became sought after as venues of choice for knowledge dissemination. With so many ambitions and interests invested in them, parliaments of science offer a valuable perspective on the complex process of urban identity formation in the Victorian era.

---

58 J.M. Ellis, 'Regional and County Centres, 1700–1840', in P. Clark (ed.), *CUHB II*, p.703.

[illegible] power in this book, however, were ones of [illegible] [illegible] through [illegible] change and [illegible] of [illegible] and communications [illegible] their effects [illegible] on the [illegible] size, power and [illegible] county [illegible] county, and [illegible] [illegible] Railways were bringing [illegible] towns [illegible] consolidating [illegible] [illegible] the widening gap between the [illegible] Manchester, Birmingham and Liverpool and the [illegible] towns like [illegible] and [illegible]. As new [illegible] industrial [illegible] [illegible]

[illegible]

[illegible] in the [illegible] from the point of view [illegible] and [illegible] their lecture halls and [illegible] [illegible]

[illegible]

[illegible] Reformation and County Towns, 1500–1800 [illegible] in P. Clark (ed.), [illegible] p. 103.

Chapter 1

Movable Feasts: Victorian Knowledge Associations and the Evolution of the Annual Meeting

The British Association, RASE, the Social Science Association and the Archaeological Institute between them held annual meetings in 62 different towns in England, Ireland, Scotland and Wales from 1831 to 1884.[1] The yearly congresses of these organisations, though similar in form, each fulfilled a slightly different purpose. What served for the Archaeological Institute as a relatively socially, if not intellectually, exclusive members' reunion was, to many in the British Association, an opportunity for proselytising to new audiences. Where the Agricultural Society sought to engage members of the farming community with new agricultural practices, the social scientists' annual meetings were about gathering together in a location where they could investigate the causes and remedies of social ills. The types of towns and the geographical locations they favoured thus differed between the four associations and over time, but the commitment of each to the principle of moving their meetings around from place to place remained a constant and was reflected in the organisational structures and working practices of the four bodies. This chapter offers an overview of the priorities of the four associations when the time came, each year, to choose a town in which to stage these flagship events, and in doing so it momentarily propels the knowledge organisations into the foreground of the study, rather than the meetings and the places which hosted them. But the need for some assessment of the perspective of the parent associations on these events is an essential precursor to the study of the meetings themselves. There was, it will be argued, a consciousness in the hierarchy of the four associations, that they were employing common means to achieve their goals of knowledge promotion and especially of the influence of the British Association for the Advancement of Science in establishing a kind of model which other national organisations could follow. By comparing the meeting patterns of the four knowledge organisations, however, it will also be shown that these annual events were adapted and utilised slightly differently in each case and that, over time, the meeting patterns of the four evolved and diverged.

The British Association for the Advancement of Science, founded in York in 1831, aimed to provide leadership to the provincial science institutions that had been springing up around the country since the latter years of the eighteenth

---

1 For a full list of the towns visited, see Appendix 1, p. 165–66.

century and to raise the profile of science nationally. Its early membership was dominated by a distinctive group of men described by Morrell and Thackray as consisting 'primarily of liberal Anglicans who possessed secure status, income, and property'.[2] As far as their scientific interests were concerned, it was geology and natural history which loomed large, as was the case in the provincial literary and scientific societies by mid-century.[3] The rationale of the BAAS, however, was to attract a wide membership open to anyone with a zest for science, and not to distinguish between teachers and students.[4] Rather than simply providing a forum in which scientific enthusiasts could meet and exchange ideas, the Association had more ambitious aims from the outset, articulated by its founding father, William Harcourt. Harcourt was an Oxford-educated cleric, geologist and philanthropist, canon of York Minster and first president of the Yorkshire Philosophical Society.[5] In the BAAS he envisaged a body which could give 'systematic direction' to scientific research and which could wield influence with the government over scientific matters.[6] The idea of using rotating yearly meetings, where scientific ideas could be discussed and debated annually in different parts of the country, did not originate with Harcourt and the other founders of the BAAS, but rather with an organisation already in existence in Germany, the *Deutscher Naturforscher Versammlung*, which had been holding scientific meetings in provincial German towns and cities since 1822.[7] Various elements of its organisational structures and methods were copied by the BAAS. For example, the arrangement of its business into sections dealing with different branches of science such as botany, zoology and mineralogy, undertaken at the second meeting in Oxford in 1832, was inherited from the German association,[8] and the latter's use of week-long provincial meetings as the main feature of its annual calendar served as a useful blueprint for the BAAS whose organisers pored over programmes and tickets issued at the German meetings for an idea of format of the annual gatherings.[9] In turn, the BAAS became a model for other knowledge associations founded in Britain in subsequent decades.

---

2 Morrell and Thackray, *Gentlemen of Science. Early Years*, p. 101.

3 Russell, *Science and Social Change*, p. 181.

4 Mumford Jones and Bernard Cohen (eds), *Science Before Darwin*, p. 298.

5 Jack Morrell, 'Harcourt, William Venables Vernon (1789–1871)', *Oxford Dictionary of National Biography* (Oxford, 2004), accessed 6 July 2009.

6 William Harcourt to Charles Babbage, 27 August 1831, quoted in Jack Morrell and Arnold Thackray (eds), *Gentlemen of Science. Early Correspondence of the British Association for the Advancement of Science* (London, 1984), p. 46.

7 Howarth, *The British Association*, pp. 6–12.

8 Archibald Geikie, *Life of Sir Roderick Murchison*, vol. 1 (London, 1875), p. 204; Howarth, *The British Association*, p. 12.

9 Bodleian Library, Dep BAAS 1, Correspondence of John Phillips (foundation volume), John Robinson to John Phillips, 25 March, 1831.

The Archaeological Institute of Great Britain and Ireland, dating from 1843 in its original incarnation as the British Archaeological Association,[10] aimed 'to popularise the study of ancient Art, to extend the field of observation, and increase the numbers of persons who take an interest in the science', and to 'preserve from destruction many valuable and beautiful specimens'.[11] The impetus for the foundation of this body lay, in part, in the inertia of the Society of Antiquaries and its uninspiring London meetings. One of its members described the 'owl-like solemnity of the scanty conclave ordinarily assembled on the appointed Thursdays at Somerset House', and the 'ponderosity or triviality of the papers'.[12] It was increasingly out of step with the numerous county archaeological societies and printing clubs springing up in British towns in the early years of the nineteenth century, and with developments on the Continent where the French Society of Antiquaries was operating a system of country-wide correspondents and an annual congress 'in some principal city or other place distinguished for archaeological interest'.[13] Closer to home, the British Association was also influential. The annual meetings of the Archaeological Institute were acknowledged as serving 'a position in the archaeological world analogous to those of the British Association in the scientific'[14] and, in this respect, the BAAS provided a 'clear model' for the newly formed institution.[15] Like the older association it arranged its business at annual meetings into sections, in this case for early and medieval antiquities, historical, architectural and archaeological studies. Notwithstanding its organisational debt to the British Association, there was some overlap and rivalry with the older organisation as far as RAI members were concerned. As early as 1856, Lord Talbot de Malahide opened the Institute's annual meeting in Edinburgh with the observation that although they felt 'the most friendly feeling towards their sister Societies, particularly the British Association ... they still could not but feel that the Association occupied some ground which properly belonged to the Institute', namely philology and ethnography, and he went on to suggest that even metallurgy and mining were matters more properly connected with the science of archaeology.[16] It was a comment which belied the anxieties of association leaders

---

10 A row after the first annual meeting at Canterbury over the publication of proceedings precipitated a split into two separate organisations. See David Wetherall, 'From Canterbury to Winchester: the foundation of the Institute', in Blaise Vyner (ed.), *Building on the Past. Papers celebrating 150 years of the Royal Archaeological Institute* (London, 1994), pp. 8–21.

11 Talbot de Malahide, quoted in Ebbatson, thesis, p. 12.

12 James Robinson Planché, *Recollections and Reflections* (London, new and revised ed., 1901), pp. 302–3.

13 Planché, p. 302.

14 SAL, WAY, SAL/MS/700/VII, Notebook, containing *The Times* obituary of Albert Way.

15 Wetherall, 'From Canterbury to Winchester', p. 11.

16 *Caledonian Mercury*, 23 July 1856.

to carve out their own distinctive knowledge niche, at a time when disciplinary boundaries were relatively undefined.

In terms of its operations, the Archaeological Institute was more London-centric than the BAAS and there was a heavy geographic concentration of members in the south and east of England.[17] There was some attempt, at the outset, to make the management of the Institute more geographically inclusive when it was agreed that 'the names of twelve or more gentlemen, not resident in London shall be added to the list of the central committee and that they shall be considered as associated with that body as honorary members', and that a number of 'local secretaries' should be nominated who could act as agents for the association in their own vicinities.[18] Despite these efforts, the membership profile of the Institute continued to reflect a south-east England bias, with London and Middlesex-based members rarely accounting for less than one-third of the total membership in the first 100 years of its operation.[19] Socially, too, its range was limited. Prominent among its ranks were clerics with an interest in ecclesiastical architecture, but it was illustrative of the broad appeal of archaeology in the mid-nineteenth century that it also attracted artists and architects, including central figures in the fashionable gothic revival of the day such as William Burges. There was a contingent from the world of publishing, including Thomas Taylor, editor of *Punch*, and civil and mining engineers were also an important presence and a number of Cambridge academics and British museum staff who also leant their support.[20] Among its most influential supporters were noblemen like the Marquis of Northampton and the Marquis of Bute who presided at annual meetings.[21] The granting of the royal charter in 1866, which allowed the Institute to change its name to the Royal Archaeological Institute of Great Britain and Ireland, came about thanks to the efforts of another influential supporter, Lord Campden.[22]

The Royal Agricultural Society of England had some similarly influential figures among its early enthusiasts, chiefly Earl Spencer and William Shaw,[23] who were crucial to the Society's formation. Along with other early activists such as Phillip

---

[17] Philippa Levine, *The Amateur and the Professional. Antiquarians, Historians and Archaeologists in Victorian England, 1838–1886* (Cambridge, 1986), p. 55.

[18] Society of Antiquaries Library, WAY, SAL/MS/700/VII, Meeting of the Central Committee, 26 March 1845.

[19] Ebbatson, thesis, p. 265.

[20] Ebbatson, thesis, pp. 25–34.

[21] Wetherall, 'From Canterbury to Winchester', p. 19.

[22] Joan Evans, 'The Royal Archaeological Institute: a retrospect', *Archaeological Journal*, 106 (1951 for 1949): 1–11. For the sake of consistency however, the acronym RAI is used throughout this book.

[23] Shaw was founder member of the English Agricultural Society and its first secretary. See Ernest Clarke, 'Shaw, William (1797–1853)', Rev. N. Goddard, *Oxford Dictionary of National Biography* (online edition, Oxford, 2004) accessed 31 July 2007.

Pusey and J.C. Morton,[24] its founders were moneyed, university-educated men and, in the case of Pusey, [25] practising farmers whose ambition was to disseminate knowledge of good agricultural techniques thereby improving productivity and prosperity throughout the country. The formation of a society which could move around the country promoting the use of innovative techniques and up-to-date tools and implements was central to this ambition, and also influential in shaping their vision were the examples of agricultural associations elsewhere in Europe such as the Paris Central Society and the Highland Society.[26] Originally known as the English Agricultural Society, RASE was established in 1838 as a kind of successor to the old Board of Agriculture. Like the other associations featured in this study, its emergence owed much to developments in the provinces where growing interest in agricultural improvement from the late eighteenth century had manifested itself in the establishment of a number of local agricultural societies. But also influential was the example of the British Association. Indeed the geologist and BAAS member Roderick Murchison, speaking at the dinner in Gloucester during show week in 1853, described the BAAS as the 'originator' of RASE.[27]

Although contemporaries like Murchison viewed RASE as an integral part of the evolving picture of scientific knowledge in the Victorian period, agricultural science has tended not to feature much in this story. Compared to historical interest in the flourishing Victorian sciences of geology, natural history and, to a lesser extent, social science, RASE and its shows have not attracted much interest from historians of science.[28] This was despite the fact that the improvement of agricultural production was becoming more institutionalised in the mid-nineteenth century, with the growth of veterinary science and the establishment of a college of agriculture in 1845.[29] The application of science to agriculture was beginning to attract the interest of some key scientific figures like Augustus Voelcker and Justus von Leibig by the 1830s and 1840s.[30] Science was central to the work of RASE from its foundation, and pursued through initiatives such as the appointment of a

---

24 Morton was a writer on agriculture and founding editor of the *Agricultural Gazette*. See Brian T. Bunting, 'Morton, John Chalmers (1821–1881)', *Oxford Dictionary of National Biography* (online edition, Oxford, 2004) accessed 31 July 2007.

25 Francis M.L. Thompson, 'Pusey, Philip (1799–1855)', *Oxford Dictionary of National Biography* (Oxford, 2004) accessed 26 July 2006.

26 Goddard, *Harvests of Change*, p. 17.

27 *Morning Chronicle*, 15 July 1853.

28 See for example, Bernard Lightman (ed.), *Victorian Science in Context* (Chicago, 1997); Daunton (ed.), *The Organisation of Knowledge*; David Clifford, Alex Warwick, Elisabeth Wadge and Martin Willis (eds), *Repositioning Victorian Sciences. Shifting Centres in Nineteenth-Century Scientific Thinking* (London, 2006).

29 J.D. Sykes, 'Agriculture and Science', in Gordon E. Mingay, *The Victorian Countryside*, (London, 1981), pp. 260–272.

30 See for example, Mariko Ogawa, 'Leibig and the Royal Agricultural Society of England meeting at Bristol, 1842', *Ambix*, 85/2 (2008): 136–52.

consulting chemist and the running of an experimental farm, and leading figures in the early years of RASE, notably Philip Pusey, were recognised pioneers of agricultural science and well connected in wider scientific circles.[31] Agricultural science did not have the broad-based appeal of, say, fossil hunting, shell or fern collecting, which provided more popular outlets for the scientific interests of the wealthy urban middle classes and the gentry,[32] but RASE certainly had serious intellectual ambitions and it approached the business of agricultural improvement with scientific rigour, aiming to 'embody such information contained in agricultural publications and in other scientific works ... to correspond with agricultural, horticultural and other scientific societies ... [and] to encourage men of science to exert themselves in the improvement of agricultural implements'.[33] It was a resolve echoed in the society's motto, 'Practice with Science'.

RASE was already well into its second decade of operation by the time the National Association for the Promotion of Social Science was formed in 1856 at the home of Lord Brougham in London's Grafton Street. A successor to earlier associations including the Law Amendment Society and the Reformatory Union, its aim was to unite the disparate groups and individuals working towards social and moral reform and, although relatively short-lived, it was a highly influential body.[34] Like the British Association, it organised its various areas of interest into 'sections'. These included jurisprudence and law amendment, education, punishment and reformation, public health and social economy. It also adopted the BAAS method of rotating its annual meetings around the country. Its founders had the British Association model firmly in mind from the outset. At the first meeting in Birmingham in 1857, Lord Brougham expressed the hope that, 'The association which we are met here to open may bear as wide a relation to moral and political science as the British Association which has now been in successful action for more than a quarter of a century, does to mathematical and physical science'.[35] In practice there were some key characteristics distinguishing the NAPSS from the BAAS, perhaps the most important of which was the close association of the NAPSS with Liberal politics. Some three-quarters of the MPs on its governing council were Liberals and the Association's historian has noted the extent to which its interests and fortunes reflected those of the Liberal Party nationally.[36] A further distinguishing feature was its greater preoccupation with social enquiry and with

---

31 Goddard, *Harvests of Change*, p. 26; Thompson, 'Pusey, Philip (1799–1855)'.

32 David Elliston Allen (ed.), *Naturalists and Society. The Culture of Natural History in Britain, 1700–1900* (Aldershot, 2001).

33 MERL, SR RASE, B1/10 Minutes of the Committee of Management (Fair Copy), 12 May 1838.

34 Goldman, *Science, Reform and Politics*, p. 29.

35 MCL, BR F306 F2, Social Science Association, Manchester Meeting, cutting from *Manchester Courier*, 25 September 1879.

36 Lawrence Goldman, 'The Social Science Association, 1857–1886: a context for mid-Victorian Liberalism', *English Historical Review*, 101/398 (1986): 56–134.

the key public health and social welfare considerations of the day. Given this focus, many of its meetings were held in rapidly rising centres of industry where social questions could be confronted first hand. But such distinctions were not always obvious to contemporaries, and the two bodies were sometimes confused. During a town council meeting in Dundee in 1862, for example, Mr James spoke about the 'very important influence' of the British Association and the way in which its deliberations on methods of suppressing crime and social evils 'very often gave rise to legislation'. He was corrected by the Dean of Guild who pointed out that 'it was actually the Social Science Association to which Mr James referred'.[37] Given the organisational similarities and a degree of overlap in membership,[38] it was an understandable mistake which, if anything, served to underline the collective view which contemporaries had of the peripatetic knowledge associations in the mid-nineteenth century.

The British Association's inaugural meeting, and thus the first British 'parliament of science', was held in York, home to the renowned Yorkshire Philosophical Society which had attracted the attention of leading geologists like William Buckland since 1821 after the discovery of prehistoric bones in Kirkdale Cave.[39] With these credentials, York was an obvious draw for the 'gentleman of science' of the British Association, interested predominantly in geology and the physical sciences,[40] but it was Harcourt who suggested that, in subsequent years, other towns and cities should be visited. In a letter to Reverend William Whewell,[41] he wrote that, 'It appears to me that the moving of such meetings from place to place would have a powerful effect in stimulating the science of those places'.[42] Press and public in the 1830s caught on to the idea quickly. The *Newcastle Courant*, reporting on the decision of the BAAS to hold its 1838 meeting in Newcastle, gave a lengthy explanation of the operation and benefits of peripatetic meetings:

> The meetings are not restricted to one vicinity. No great metropolis swallows all the good that is derivable from them by those among whom they take place. The members proceed from town to town; and thereby not only afford themselves fresh fields for observation in every new locality, but inspire with a taste for scientific enquiries the several communities in which, for a time, they reside.[43]

---

37 The *Dundee Courier and Argus*, 29 August 1862.

38 Ronald K. Huch, 'The National Association for the Promotion of Social Science: its contribution to Victorian health reform, 1857–1886', *Albion*, 17/3 (1985): 298.

39 Orange, 'Science in early nineteenth-century York', pp. 6–7.

40 Morrell and Thackray, *Gentlemen of Science. Early Years*, p. 101.

41 Whewell was professor of mineralogy and later of moral philosophy at Cambridge, and Master of Trinity College Cambridge from 1841 to 1866. See Morrell and Thackray, *Gentlemen of Science. Early Correspondence*, p. 13.

42 Morrell and Thackray, *Gentlemen of Science. Early Years*, p. 81.

43 *Newcastle Courant*, 22 September 1837.

But the itinerant meetings model was not just devised for the scientific benefit of the population. The rotating annual gatherings ranging across provincial Britain gave these organisations their claim to nationwide reach at a time when science was sensitive to international comparisons and aware of the deficiencies in some of its older institutions. The nation was 'a vehicle through which the political aspirations of scientific communities and science's claims to public usefulness could be articulated and realised'.[44] In other words, the roving annual meetings were a key legitimising tool, designed to extend the influence and enhance the reputation of the associations which used them.

The roaming annual meetings pioneered by the BAAS were week-long affairs, often running from Wednesday to Wednesday. Commencing with an address by the president for the year, the meeting then divided into its various 'sections' where papers on specialist scientific topics were delivered in separate, parallel sessions. Attendance figures fluctuated from year to year. In 1869, for example, Exeter organisers were satisfied that the 1,856 tickets issued for the BAAS meeting compared favourably with delegate numbers for meetings held in other, similarly sized towns.[45] The Saturday of meeting week was often set aside for excursions to local places of scientific interest and, as we shall see in Chapter 3, evenings were utilised for a combination of social and scientific purposes, with popular lectures, soirées and dining occupying much of the delegates' time.

This meeting model, tried and tested by the BAAS since 1831, was easily adaptable for use by the archaeology and social science associations which were similarly structured around sections, but the Royal Agricultural Society had rather different needs. Agricultural shows, organised locally and regionally by English and Welsh county agricultural societies, and by the Highland Society in Scotland, were already an important fixture on the provincial farming calendar by the 1830s. A 'more elaborate version of the traditional local fair', these shows involved the awarding of prizes for livestock and were open to the public.[46] RASE's first annual show, or country meeting, which was held in Oxford in 1839, drew on these traditions, but also on the example of the British Association and, crucially, on the availability of railway transport to introduce the principle of moving annual agricultural shows for the first time. These week-long events, which soon became the centrepiece of the society's year, incorporated a wider range of activities than the other annual meetings under scrutiny here, and included the showing of cattle, trials of agricultural implements, exhibitions and a grand banquet. Attendance figures ran into the tens, and occasionally the hundreds, of thousands as the society sought to establish a base in a populous town from where the sizeable agricultural population of the rural hinterland could also be attracted. Crowds like the 62,539 who attended the 1858 show in Chester and the 90,011 who flocked to Carlisle

---

44 Ludmilla Jordanova, 'Science and Nationhood: cultures of imagined communities', in Geoffrey Cubitt (ed.), *Imagining Nations* (Manchester, 1998), p. 195.

45 *Devon Weekly Times*, 20 August 1869.

46 Hudson, *Patriotism with Profit*, pp. 53–7.

on the occasion of the society's meeting there in 1880 were not exceptional.[47] As these figures indicated, RASE meetings were mass participation events. The language used to describe RASE events was unlike the others featured in this study. They were sometimes described as 'meetings' but often as 'shows', 'festivals' or even 'great encampments'[48] and their size and attendance levels made them much more easily distinguishable by contemporaries from the meetings of the BAAS. One newspaper report drew the distinction by explaining that, 'The latter travels mainly for the instruction and gratification of its members ... [and] the Royal Agricultural ... seeks the countenance of every person who will favour the show yard with their presence'.[49]

The Social Science Association's first meeting took place in Birmingham in 1857 and other early venues included Bradford, Liverpool and Glasgow. Its events were much more like those of the British Association, with a week-long programme of lectures and discussion organised around a key opening address followed by a series of specialist sectional meetings, but there were also some

Figure 1.1 The Royal Agricultural Society's showground in Wolverhampton in 1871, *Illustrated London News*, 8 July 1871 (author's collection)

47 Nicholas P.W. Goddard, 'The Royal Agricultural Society of England and Agricultural Progress, 1838–1880', unpublished Ph.D. thesis (University of Kent, 1981), p. 314.

48 For examples of the different descriptions of RASE's annual events, see for example, *The Derby Mercury*, 19 July 1843, 'The General Meeting'; *The Morning Post*, 19 July 1854, 'The anniversary festival and show'; *The Times*, 21 July 1876, 'the annual great encampment'.

49 *Nottinghamshire Guardian*, 5 April 1867.

important distinctions between BAAS and NAPSS events. The Social Science Association encouraged a much greater level of women's participation both as delegates and platform speakers at these meetings. At the Liverpool meeting in 1858, for example, some 596 ladies' tickets were sold out of an estimated total of 2,000 in attendance.[50] Prominent female contributors such as Mary Carpenter, Louisa Twining and Frances Power Cobbe took part in debates on juvenile delinquency and the Poor Law.[51] Working-class involvement was also an important objective of the Social Science Association. Special lectures for working men were organised as an integral part of the annual meeting programmes and 'by a rule of the Association any Mechanics Institute, Working Mens' College or other similar body, can become a subscribing member on payment of one guinea a year, with the privilege of sending two representatives (working men) to the meetings of the Association'.[52]

Such mechanisms designed to promote engagement with the working classes did not feature in the annual meeting arrangements of the Royal Archaeological Institute. Their gatherings were smaller in scale and less about proselytising to the widest possible audience. Instead, the annual meeting was designed to serve the intellectual and social interests of the existing membership rather than to extend the boundaries or remit of the organisation. Some 200–300 people attended the archaeologists' first gathering at Canterbury in September 1844,[53] and thereafter there was a tendency on the part of some participants to view annual congresses, and especially the social events that took place during meeting weeks, as opportunities for reunions of existing members rather than for extending the fold to include new participants.[54] These events were undeniably very different from those of the BAAS, NAPSS and RASE. Compared with the meetings of those organisations, the modest scale of RAI gatherings might have limited their kudos and commercial potential, but in other respects they had equal or greater appeal to prospective host towns. More than the other associations, the Archaeological Institute drew attention to the historic assets of the localities in which their meetings were held, highlighting the importance of ancient monuments, churches and archaeological sites. The small scale, scholarly atmosphere of the annual meetings of the RAI left them less open to criticism in the contemporary press than some of the other associations. 'Scholarlike speeches' and 'able papers' were the typical bill of fare

---

50 NML MAL, John Towne Danson, D/D/111/17/5a, Newspaper Cutting Book, *Northern Times*, 15 November 1858.

51 O.R. McGregor, 'Social Research and Social Policy in the Nineteenth Century', *British Journal of Sociology*, 8 (1957): 153.

52 UCL Special Collections, Brougham MSS, 13,151, G.W. Hastings to Lord Brougham, 1858.

53 Planché, p. 303.

54 See for example the comments of the Marchioness of Ripon in 1874, SAL, WAY MS/700/VII, unidentified newspaper cutting, possibly from *The Times*.

served up at the Institute's annual congresses,[55] and by the late 1860s, despite the continued operation of rival archaeological bodies, they were acknowledged in some quarters at least as 'not only the leading but as the most promising of all our antiquarian societies'.[56]

The yearly gatherings of members in a provincial town or city quickly became the mainstay of the calendar of activities employed by Victorian knowledge associations. In reference to the NAPSS, the editors of the *Manchester Guardian*, writing in 1879, observed that:

> The Association for the Promotion of Social Science is the stem of which the annual congress is the flower, but we shall probably not be far wrong in assuming that the flower is incomparably the more important part of the organisation and that the body to which it belongs would be nothing without it. The congress is in fact the Association ...[57]

Their comment could equally have been made of the other three organisations featured in this study. While their monthly council meetings were reported in the press, and their journals and transactions published and distributed to members, the annual meeting was the only time that the existence and activities of the BAAS, RASE, NAPSS and the RAI came to the attention of the wider public. The meetings also had the potential to generate funds which could be used to finance other aspects of the associations' work. The NAPSS' 1863 meeting in Edinburgh yielded a sum of £1,500 which enabled the Association to establish a permanent office in London from where work on subjects of special interest could continue between annual meetings. By the same period, the British Association had also built up a reserve fund from annual meeting revenues which enabled it to spend around £3,000 per year on various scientific projects.[58]

The importance of the annual congress to the Victorian knowledge associations was such that they were sensitive to any criticism levelled against their flagship events. In a period when, according to one commentator, the 'boundaries of scientific legitimacy' were being debated,[59] some contemporaries expressed concern that large, popular meetings were not necessarily the best medium for holding intellectual debate and discussing scientific ideas.[60] Among RASE members, doubt was expressed in some quarters of the society about the degree of educational impact it was having on agricultural districts. Some members tried

---

55 *Bristol Mercury*, 2 August 1851.

56 *The Standard*, 30 July 1868.

57 MCL, NAPSS, BR F306 S2, cutting from, *Manchester Examiner*, 1 October 1879.

58 MCL, BR F306 S3, Social Science Association, Manchester Meeting, 1866, extract from *Manchester Guardian*, 4 October 1866.

59 James G. Paradis, 'Satire and Science in Victorian Culture', in Bernard Lightman (ed.), *Victorian Science in Context* (Chicago, 1997), p. 146.

60 See, for example, Wetherall, 'From Canterbury to Winchester', p. 12.

to anticipate and deflect possible criticism. J.C. Morton, in particular, did not believe that the holding of an annual show and the publication of the society's journal were enough to stimulate improvement in agricultural education and he said so at the society's general meeting in 1863. As a result, a new education committee was established, and a prize essay competition inaugurated at the 1865 show in Plymouth.[61] Compared with the other associations, it was the often lavish and showy annual meetings of the British Association that attracted the highest levels of press criticism. The BAAS was lambasted as early as 1835 for displaying greater interest in dining and hospitality than in scientific debate at its meeting in Dublin. It was possible, according to one worried member, to detect 'a certain taint of ridicule beginning to be attached to the proceedings in the eye of the public'.[62] It was not just the levels of hospitality on offer at meetings which attracted scorn. Contemporaries made judgements on the quality of knowledge on offer and were particularly quick to deride evidence of superficiality. NAPSS meetings, although they attracted the leading social reformers of the day and in some cases had a direct influence on social legislation and policy-making,[63] also faced the charge, 'that we are merely a meeting where people talk and when the meeting is over and the talk is at an end, no further good results from it'.[64] The editors of *The Times* queried the worth of the papers presented at the early congresses of the Social Science Association, arguing that, 'There is no subject on which the most erratic essayist could not claim to be heard'.[65] After the Edinburgh meeting in 1863, one newspaper report attacked what it called 'these touch-and-go congresses', for tackling numerous disparate issues but making meaningful progress with none of them. 'Twenty or thirty subjects per diem', it suggested, was the average ground covered, making the event 'a great holiday for hobby-horse riders'.[66] For an organisation which aimed to influence government policy on education, law reform, health and sanitation, this was serious censure, and thus initiatives to continue the work of social science in the provinces between annual meetings were enthusiastically supported. Local associations were spawned in a number of the Association's early meeting places, including Bradford, Liverpool, Edinburgh and Aberdeen.[67]

Given the levels of scrutiny directed at annual meetings, they became the main events on which the energies of the associations were focused, and around which

---

61 Goddard, thesis, pp. 441–6.

62 Bodleian, Dep. BAAS 59, John Robison (Edinburgh) to John Phillips, 23 September 1835.

63 Goldman, *Science, Reform, and Politics*, pp. 262–89.

64 NML MAL, John Towne Danson, D/DIII/15/3, National Association for the Promotion of Social Science, printed pamphlet: Proceedings of the meeting of the General Committee of the NAPSS on 5 May 1858.

65 *The Times*, 26 September 1860.

66 *Lloyd's Weekly Newspaper*, 18 October 1863.

67 Goldman, *Science, Reform and Politics*, p. 86.

their strategies and operational procedures were arranged. The associations went to great lengths to ensure that everything possible was done to make them a success. Fundamental to this process was the matter of finding the right environment in which to hold the congress. In the early 1830s, when the idea of holding mass learned gatherings was still novel, large towns and cities were more readily linked with social problems and unrest than they were with intellectual exchange. Such perceptions of city life prevailed not only in British but also in American culture.[68] Unrest over political reform and fears of cholera cast a shadow over social relations in a number of British cities and several voices were raised in favour of a postponement of the inaugural meeting of the British Association for the Advancement of Science in 1831 because of 'the present excited state of feeling on other subjects'.[69] The context for this concern was the reform agitation of that year, culminating in serious rioting in some of the country's largest provincial towns, such as Bristol.[70] The Association's historians have pointed out that, while the choice of meeting places did not directly correlate with the geography of social unrest, this was a factor in deciding suitable venues.[71] They briefly considered postponing the annual meeting organised in Birmingham in the summer of 1839 because of 'the disturbed state of the town' which was the scene of Chartist agitation.[72] The safety and health of the urban environment featured in discussions on meeting arrangements for other learned associations too, but organisers were often inclined to set aside their concerns for the sake of allowing the annual meeting to go ahead. Worries about repeal agitation, for example, were overlooked by the BAAS for their visit to Cork in 1843,[73] and the outbreak of cholera in 1849 did not prevent the holding of the usual range of summer congresses that year, although it affected attendance levels at the Archaeological Institute's meeting in Salisbury, causing people to stay away for 'dread of the prevailing epidemic'.[74]

Other circumstances beyond the control of the associations also occasionally disrupted meeting plans, such as the outbreak of cattle plague which forced the abandonment of the planned RASE show in Bury St Edmunds in 1866.[75] But the associations did everything they could to find unproblematic dates and venues. This involved avoiding calendar clashes with the activities of other learned societies during the increasingly crowded months of summer and early autumn. The NAPSS, for example, tried to avoid choosing the same meeting week as the BAAS and took into account the prospect of an election and the dates of foreign

68 Bender, *Intellect and Public Life*, p. x.

69 Bodleian Library, Dep. BAAS 1, John Robison to John Phillips, 25 March 1831.

70 Mark Harrison, *Crowds and History. Mass Phenomena in English Towns, 1790–1835* (Cambridge, 1988), p. 147.

71 Morrell and Thackray, *Gentlemen of Science. Early Years*, p. 106.

72 Bodleian Library, Dep. BAAS 17, 31 July 1839.

73 *The Times*, 18 August 1843.

74 *Hampshire Advertiser* and *Salisbury Guardian*, 28 July 1849.

75 Goddard, *Harvests of Change*, p. 35.

congresses which might be of interest to its members. A further consideration was the impact of local events on the availability of visitor accommodation in the town. The scheduling of their Sheffield meeting in 1865 proved problematic on all of these fronts, as George Hastings, the Association's, secretary explained to Lord Brougham:

> With regard to the time of our Sheffield meeting, it is certain that September is an impossible month. The British Association meets on the 6th of that month, and after that comes Doncaster races, and then a great Volunteer gathering near Sheffield. So the choice is between October and August. Each has its advantages and disadvantages which we must balance as well as we can. I want to find out when the dissolution is likely to take place, as whatever we do we must avoid that. Also I have a letter this day from Corr Van der Maesner of Brussels saying that the International Social Science Association will meet at Berne in the last week of August ...[76]

The NAPSS settled on late September or early October as the usual time for its congresses, thereby waiting until after the other associations' events were over.[77] But it still felt the effect of the others, the British Association in particular, on its revenues and membership. Hastings claimed that: 'we are hit hard by the British Association which draws off in money, men and papers, much support that we should otherwise obtain'.[78]

Settling the question of when to hold the annual meeting, however, was relatively simple when compared to the more thorny issue of location. Once established, the congresses became sought-after events and associations found themselves in receipt of multiple invitations from aspiring host towns. Methods had to be devised to vet and select an appropriate venue. As far as the BAAS was concerned, delegations from any town or city hoping to stage a meeting could attend the annual council to put their case to the association hierarchy of the BAAS for consideration. Before dispatching their delegations, many first sought advice from John Phillips, the Association's general secretary. John Jeffray, honorary secretary of the Birmingham and Midland Institute, wrote to Phillips in June 1860 as arrangements were being made to send a Birmingham delegation to the Oxford meeting stating that, 'We propose to ask that Birmingham may be selected in 1862. Can you give us any hints that guide our movements?'[79] Phillips played a crucial

---

76 UCL Special Collections, B MSS, 13,146, George W. Hastings to Lord Brougham, 3 March 1865.

77 The one significant exception was the London congress, held in June 1862. Dates for each of the NAPSS congresses can be found in Goldman, *Science, Reform and Politics*, p. 382.

78 UCL Special Collections, B MSS, 35,214, George W. Hastings to Lord Brougham, 23 December 1861.

79 Bodleian Library, Dep. BAAS 59, John Jeffray to John Phillips, 15 June 1860.

role in communications between the BAAS' executive council and their would-be hosts in the towns. As well as corresponding with contacts like Jeffray from an early stage in the process, he also directed arrangements once a meeting venue had been announced and, in 1833, the Association increased his annual salary to £200, 'to enable him to attend at the places of meeting for the purpose of making arrangements previous to the assembly of the association and to bear such expenses as the council may think proper to indicate'.[80] In the other associations, too, the role of secretary was pivotal to the success of annual meetings and the key point of contact between the central administration and organisers in the host towns. The Archaeological Institute in its formative years owed a great debt to the energy and vision of Albert Way, who was honorary secretary of the Institute from 1845 to 1868. He edited its publication, the *Archaeological Journal* and was a leading light in promoting 'a more scientific approach' to archaeology.[81] In the Social Science Association, George Woodyatt Hastings fulfilled the role of secretary for the first decade. A practising barrister and previously secretary for the Law Amendment Society and the National Reformatory Union, Hastings' expertise in matters of social and legislative reform, combined with his connections in contemporary political circles, made him an ideal choice.[82] In RASE the post of secretary changed hands three times between 1838 and the mid-1880s, with H.M. Jenkins acknowledged as perhaps the most successful incumbent, combining the post with editorship of the *Journal* from 1868 to 1887.[83]

RASE's salaried officers had to be effective. The Society operated a far more systematic and formal system of selecting places in which to stage its annual show than did the other associations under scrutiny here, and it had to be carried out with care, tact and rigour if towns were to engage in the process. First suggested by William Shaw in 1840, the Society's method was to divide up England and Wales into districts and invite applications to host from towns in a different district each year according to a pre-announced schedule.[84] The aim of this was to ensure that each part of the country played host to the annual show of the Society in turn.[85] The system was not strictly adhered to. As early as the mid-1840s, RASE began to depart from its planned district rotation, deciding not to visit south Wales in 1847 because of its proximity to Shrewsbury where the 1845 meeting had been held.[86] A special council met in 1846 to adjust the regions and set a new rotation for the

---

80 Bodleian Library, Dep. BAAS 131, Notebook of John Phillips, 1833, p. 78.

81 Bernard Nurse, 'Way, Albert (1805–1874)', *Oxford Dictionary of National Biography* (online edition, Oxford, 2004), accessed 18 February 2012.

82 Lawrence Goldman, 'Hastings, George Woodyatt (1825–1917), *Oxford Dictionary of National Biography* (Oxford, 2004), accessed 4 March 2012.

83 Scott Watson, *The History of the Royal Agricultural Society*, p. 165.

84 Goddard, thesis, p. 306. As its name suggested, RASE's remit did not encompass Scotland or Ireland, but its coverage of other regions was extensive.

85 MERL, SR RASE B/I/1, Minutes of Council (rough copy), December 1841.

86 MERL, SR RASE /B/I/3, Minutes of Council (rough copy), 1 April 1846.

next five years.[87] Not all of the specified 'divisions' contained the same number of sizeable towns capable of playing host to these increasingly large-scale events. This meant that the regions had to be re-drawn a number of times and varied greatly in size according to population density and levels of urbanisation. In 1876, for instance, it was agreed that district B should consist of the single county of Yorkshire, while district F comprised, Cornwall, Devon, Somerset, Hampshire, Berkshire, Oxfordshire, Surrey, Sussex and Middlesex.[88] With the same frequency of visits to both of these vastly differing districts, the inequities of the system were obvious, but in general, as Figure 1.2 shows, the Society's annual shows, more than the meetings of the other associations, genuinely traversed the regions. By covering such a lot of ground RASE could legitimately claim to be taking its key messages on agricultural improvement far and wide around the country, but there were also drawbacks to the system. As will be seen in Chapter 2, it had the side-effect of thrusting neighbouring towns into direct competition with one another in a bid to be the host venue in any given district. The Special Country Meetings Districts Committee tried to mitigate this when it overhauled the system for selecting host towns in 1876. Its aim was to 'obviate the great waste of time and money which has hitherto been incurred by the authorities of several towns in each district in getting up subscriptions and obtaining sites for show-yard and trial lands in order to compete for the society's shows and also to prevent the disappointment felt by the inhabitants and others connected with such towns as have competed unsuccessfully'.[89] This willingness to revise the districts and keep the system of town selection under review was a sign of RASE's responsiveness to urban developments in the provinces and its ability to view the whole process from the perspective of aspiring host towns.

The BAAS did not attempt anything like the regional approach of RASE, but still achieved an impressive level of geographical coverage in its first 50 years of operation. Like its agricultural counterpart, there was a good deal of debate and discussion in the British Association about the choice of venues for annual meetings. By the 1840s, when a decade of activity had seen the BAAS exhaust most of its favourite, accessible provincial centres, voices from within the association argued for consideration of some new venues. William Whewell urged his fellow BAAS members in 1844 to think about 'Leeds, Hull, Lincoln, Norwich, Portsmouth, Exeter. You ought to go to the frontier of Wales, to Swansea or Gloucester … to Shrewsbury',[90] but Roderick Murchison replied that, 'We think that nearly all the places you enumerate are wholly incapable of receiving the BA

87 Ibid. They settled on Yorkshire for 1848, the Eastern region in 1849, the South West in 1850 and Middlesex in 1851.

88 MERL, SR RASE BXIII/4, Miscellaneous, Special Country Meeting Districts Committee, 1 October 1876.

89 Ibid.

90 William Whewell to Roderick Murchison (27 February 1844) quoted in Morrell and Thackray (eds), *Gentlemen of Science. Early Correspondence*, p. 359.

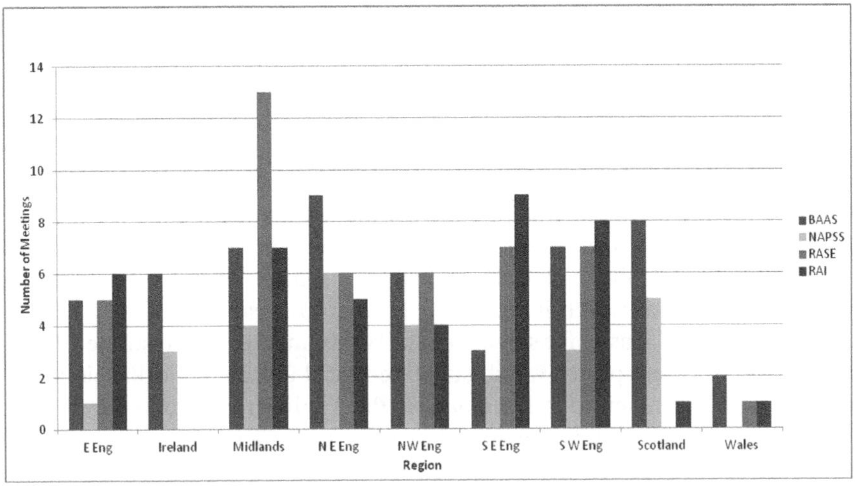

Figure 1.2 Regional distribution of meetings, 1831–84

in its present stature'. Indeed in an earlier letter he had named Hull, along with Halifax, as 'being places which I trust we shall never visit'.[91] The debate revealed the tension between those who saw the annual meeting principally as a reflection of the status and prestige of the Association, and those for whom the 'missionary' agenda behind the annual meeting outweighed any commercial or status issues. By the time these letters were being penned, the latter view was already gaining the upper hand. Hull and Swansea were both visited within a decade. The presidential address at the latter town's annual meeting in 1848 described Swansea as, 'remote from the metropolis, remote from the chief seats of English learning, remote also from the great highways of communication', but acknowledged that the Association was looking, 'to carry the torch of science everywhere'.[92]

The Royal Archaeological Institute was much less concerned with geographical coverage in its choice of meeting places. In all, between 1838 and 1886, 30 per cent of its annual meetings were in the south east and a further 27 per cent in the south west.[93] Despite the fact that its full title was the Royal Archaeological Institute of Great Britain and Ireland, it made few attempts to extend its influence beyond this south of England base. On occasions when its meetings were held in other parts of the country, they were sometimes poorly supported by key members. Edward Freeman, writing after returning from Hull where the RAI held its meeting in 1867,

---

91 Roderick Murchison to William Whewell (26 September 1840) quoted in Morrell and Thackray (eds), *Gentlemen of Science. Early Correspondence*, p. 339.

92 *Report of the Eighteenth Meeting of the British Association for the Advancement of Science*, President's Address, 9 August 1848 (London, 1849).

93 Levine, *The Amateur and the Professional*, p. 49.

noted the absence of Hook, Willis, Stubbs, Guest and 'divers other good people who did not appear'.[94] The so-called 'Celtic periphery' was particularly poorly served with annual meetings, the RAI only once venturing into Scotland in its first two decades, waiting until the 1870s before visiting Wales and neglecting to visit Ireland, although Dublin was considered as a venue in the early 1850s.[95] There was a keen awareness among members of this Anglo-centricity, but little appetite to address it. In fact, attempts were made to justify the neglect of outlying regions. Wales, it was suggested, was avoided because, 'The Principality has long supported the well-known Cambrian Archaeological Association, of the proceedings and published records of which it would be difficult to speak too highly'.[96] This, of course, was no real reason for staying out of Wales; many of the Institute's English meeting venues were places which supported local archaeological societies. The fact that Wales was viewed as lying outside the Anglican mainstream by a society which drew the core of its membership from that source is a more plausible explanation.[97] For different reasons, the Social Science Association was also less geographically adventurous than RASE or the BAAS. Keen to hold its congresses in places where the social problems of the day, and practical solutions devised to solve them, could be examined at close quarters, the NAPSS was naturally drawn to bigger centres of population. The growing industrial towns of the north of England and Scotland exerted a much greater pull than did the south and east, or Wales. Rather than holding formal competitions to select host towns, the Social Science Association's secretary simply received invitations which were then put before the committee for discussion. By the 1860s, however, when the NAPSS was gaining prominence and status, some towns began sending deputations in a similar manner to those of the BAAS, to its council to argue their case for hosting a meeting.[98]

The rather different priorities and emphases of the associations which lay behind these diverse geographical meeting patterns also shaped the associations' preferences for particular kinds of host town (Figure 1.3).[99] The first meetings of the

94 William Stubbs, Bishop of Oxford, and Walter F. Hook, Dean of Chichester, along with the architectural historian Robert Willis and the philologist, Edwin Guest, were among Freeman's friends and contacts at archaeological meetings. Edward A. Freeman to Dean Hook , 3 September 1867, quoted in William R.W. Stephens (ed.), *The Life and Letters of Edward A. Freeman*, vol. 1 (London, 1895), p. 389.

95 SAL, WAY MS/790, *Proceedings at the Meeting of the Archaeological Institute*, annual meeting, Newcastle-upon-Tyne, 1852.

96 Supplementary notice of 'Annual Meeting for 1871 to be held at Cardiff', *Archaeological Journal*, 27 (1870).

97 Ebbatson, thesis, p. 17.

98 Goldman, *Science, Reform and Politics*, p. 91.

99 The town categories used here, and also in Figure 1.4, are adapted from the 'six classes of towns in England and Wales' as defined in, PP 1863 LIII pt.1, *Census of England and Wales, 1861*. General Report, Appendix to Report, p. 105.

BAAS, NAPSS, RASE and the RAI were held, respectively, in York, Birmingham, Oxford and Canterbury. In contrast to the hard-fought competitions that were to develop, not much discussion was generated over the merits of these early venues. The presence of a group of local enthusiasts, sufficiently organised to issue a timely invitation, and willing to shoulder the burden of planning the meeting, was enough to convince the associations' leaders that their event was in safe hands.[100] These locations reflected many of the preferences which the associations were to exercise over the coming decades. Regional capitals and university towns were the favoured haunts of the BAAS in its early years. It was in these kinds of venues, mingling with key figures from the British scientific establishment, that BAAS members felt most at home.[101] The Royal Archaeological Institute had rather different priorities in mind when selecting its annual meeting venues. Cathedral towns like Hereford were a popular choice, where members could tour a twelfth-century church, view the site of a Roman encampment and receive hospitality from local dignitaries.[102] For both of these associations, concerns over the intellectual validity of their meetings could be mitigated, to some extent at least, by choosing a host venue which had an established track record as a centre for enlightened thought and learning. University towns like Dublin and Edinburgh, for instance, provided a suitable environment, as did places which had well-renowned institutions of culture and learning supported by active local elites. Thus towns with well-resourced literary and philosophical societies peopled by distinguished provincial scientific figures were more likely to be looked upon favourably as meeting places.

For RASE, there was a need to identify towns which served a large rural hinterland and which might thereby provide a base for disseminating its ideas on agricultural improvement over as wide an area as possible. The traditional county towns with their local agricultural societies and their role as market and distribution centres for the produce of the wider agricultural region made them a popular choice. Yet such places were considered too small to be suitable as venues for the NAPSS, with its interest in social enquiry and the key public health and social welfare considerations of the day. In 1864, for example, Gloucester was dismissed as a possible contender for the Social Science congress after Hastings reported to Lord Brougham that, 'I hear it is an old county town and nothing more'.[103] In contrast, Gloucester had been chosen by RASE as early as 1853. With its remit to educate the agricultural communities and its need to find places for its

100 Oxford, for instance, was chosen in 1839 by RASE, because of its centrality and because assurances of support for the meeting were received from academic and municipal authorities in the town. See MERL, SR RASE XXXII/1, Letterbook, 1838–44, 2 August 1839.

101 Morrell and Thackray, *Gentlemen of Science. Early Years*, p. 99.

102 *The Times*, 15 August 1877.

103 UCL Special Collections, Brougham MSS, George W. Hastings to Lord Brougham, 5 December 1864.

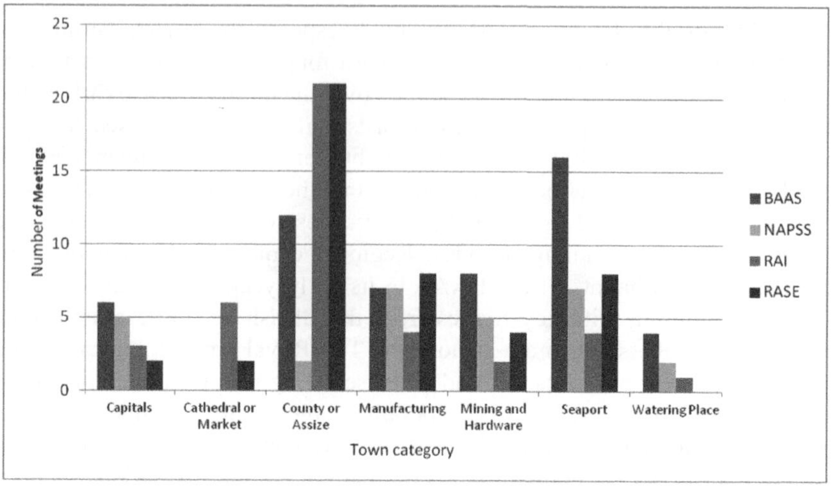

Figure 1.3 Annual meetings by town type, 1831–84

meetings which could represent and attract people in from an extensive hinterland, RASE looked much more favourably on a town with an established role as a market and supply centre for a wider farming region.[104]

As well as looking to secure a venue that would help them reach their target audience and protect the intellectual credibility of their activities, the associations also sought out meeting places where attendance levels were likely to be high. Some contemporary observers were pragmatic enough to admit that, for many delegates, the attractions of the host town exerted a more powerful pull than the content of the learned meeting. At Edinburgh, where the Social Science Association met in 1863, one newspaper correspondent wrote that, 'The place of meeting, indeed, was of itself a guarantee of success. Independent of the somewhat doubtful attractions of the learned papers and learned discussions of our social regenerators, "Edina, Scotia's darling seat", has charm of scenery, society and association, sufficient to allure a good attendance of strangers'.[105] Good attendance figures at annual congresses were probably the most important public sign of the associations' status and strength. They also generated revenue, although takings for early meetings were often quite modest. The secretary of the NAPSS estimated that 'after paying all expenses there is a surplus of £300' following the inaugural Birmingham meeting in 1857.[106] But as the stature of the associations grew, the

[104] Peter Clark, 'The civic leaders of Gloucester, 1500–1800', in Peter Clark (ed.), *The Transformation of English Provincial Towns, 1500–1800* (London, 2nd edition, 1985), p. 313.

[105] *Newcastle Courant*, 16 October 1863.

[106] UCL, Special Collections, B MSS 36,307, George W. Hastings to Lord Brougham, 3 November 1857.

scale and potential revenues from their meetings increased correspondingly and thus the choice of a venue that was easily accessible, attractive to members and capable of accommodating large numbers of delegates was a top priority. Towns with good railway links were regarded as highly desirable. Indeed, as Lord Talbot de Malahide observed at the RAI's annual meeting at Ripon in 1874, 'The use of peripatetic societies was quite a creation of modern times, and was in great measure produced by the increased facilities for locomotion produced by railways and steamboats. Without these aids they would hardly have been possible'.[107] Ventures to places regarded as geographically remote were perceived as risky and financially less rewarding for the learned associations, but were feasible as long as they were interspersed with the choice of larger, more central venues. The British Association's meeting in Cork in 1843, for example, required lengthy travel for most delegates – even those within Ireland, many of whom were concentrated around the main population centres of Dublin and Belfast. The resulting attendance of just 370 was paltry by the BAAS' usual standards, with its annual meetings often attended by three or four times that number.[108] The failure of the Cork BAAS meeting to attract the usual large attendance was believed, in some quarters of the press, to have cast doubt on the future of the association which was not removed until a more secure meeting in its birthplace, York, the following year where it 'appears to have revived … all its popular interest which was somewhat damaged by its exile last year to the remote regions of Munster'.[109]

Over the study period as a whole there was a general shift to bigger and more populous locations. This trend, which is illustrated in Figure 1.4 by the growing share of meetings held in manufacturing towns and seaports by the end of the period, reflected the desire of learned associations to stage ever bigger and more lucrative events. By 1861 there were 72 large towns in England and Wales with populations of over 20,000,[110] and, by the final third of the study period, the associations were overwhelmingly favouring these places, or Irish and Scottish towns of equivalent size, for their annual meetings. Of the 71 British-based meetings of the four featured associations, held in the 1867–84 period, only 10 were held in English towns like Lewes, Ripon, Hereford and Taunton, which were not included in the 1861 Census' 'large towns' classification. For RASE, urban locations where 'receipts were swelled by non-agricultural spectators and a sound financial result was more certain' were increasingly preferred to smaller centres where meetings might be held at a loss.[111] As a consequence, there was less incentive to locate the annual shows in places which had close links with the countryside. But at the same time, RASE council members were anxious that,

---

[107] SAL/MS/700/VII, newspaper cutting on the RAI meeting at Ripon.

[108] Figures from *Cardiff and Merthyr Guardian*, 26 August 1848.

[109] *Exeter Flying Post*, 10 October 1844.

[110] PP 1863 LIII pt.1, *Census of England and Wales, 1861*. General Report. Appendix to Report, pp. 102–3.

[111] Goddard, *Harvests of Change*, p. 36.

in their choice of annual meetings, the agricultural improvement agenda should not be seen as subordinate to the money-making potential of a populous venue. It was a difficult balance to strike but one on which the credibility of the Society rested. In 1871, council members debating the relative advantages of Wales' 'coal metropolis', Cardiff, over Cheltenham were aware of the lure of the 'pitman's shilling' but also of the possible criticism they would attract if they got into the habit of going 'about to collect shillings'.[112] But their creeping preference for more populous places had not gone unnoticed. In 1876 *The Times* speculated that 'the time may have arrived when small county towns, old cathedral cities and remote seaports can no longer be regarded as proper and profitable places in which to pitch the annual great encampment of the Royal Agricultural Society'.[113] In that year its chosen venue was Birmingham, followed by Liverpool in 1877, seemingly signalling a shift away from the modest market towns of earlier decades, but, interestingly, this trend was neither sudden nor decisive.

As Figure 1.4 also reveals, the popularity of county and assize towns as venues for annual meetings by no means collapsed. They continued to constitute the largest number of venues in the last third of the study period. A number of factors combined to ensure that various parliaments of science continued to make their appearance in a wide range of urban settings. In part it was a consequence of developments in the urban infrastructure: improved railway connections meant that, by this period, more of these towns were accessible to visitors from different parts of the country. It was also a reflection of the unchanging preferences of the RAI, which continued to adhere to the older towns and cities with cathedrals, castles or other ancient remains. Such places had an enduring appeal to other associations too. Places with an established track record as centres of culture and polite leisure had many advantages over newer, commercial and industrial towns when it came to hosting meetings, and the latter sometimes fell disappointingly short of expectations. The experience of Huddersfield, which was visited by the Social Science Association for its annual meeting in 1883, illustrated the problem. As an expanding town at the centre of an important textile region it was an attractive location for a Social Science congress, but it lacked the range of venues on offer in other centres. *The Times* reported that, 'the location does not favourably impress visitors, the majority of whom will be widely dispersed owing to the absence of accommodation near the centre of town'.[114] It was an issue which, as we shall see in the following chapters, also troubled other industrial centres with aspirations to play the role of meeting host.

As the survey offered in this chapter has shown, the evolution of the roaming annual meeting as a tool for knowledge dissemination in Victorian Britain was closely bound up with the establishment of the BAAS and its imitator associations. It was the concept of the moveable meeting which gave these organisations their

112 MERL, SR RASE/B/1/7, Minutes of Council (rough copy), 3 May 1871.

113 *The Times*, 21 July 1876.

114 *The Times*, 3 October 1883.

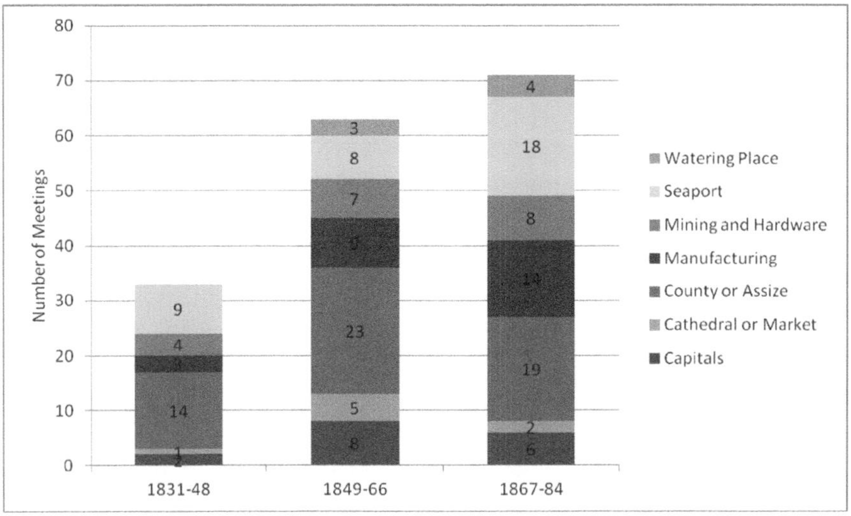

Figure 1.4 Annual meetings by town type: changing patterns over the study period

provincial platform and their claim to nationwide reach. But although they were devised as a framework in which the associations and their members could operate, and were located in places which met their criteria and priorities, the meetings were ultimately reliant for their effectiveness on the willingness of towns and cities to host them. Without this reservoir of goodwill and shared interest on the part of the towns, the whole peripatetic knowledge project would have been unworkable. In fact, as the next section of this book will show, towns were eager to play their part and came forward in ample numbers over successive years to offer themselves as meeting hosts. An insight into their motives for participating is revealed as they engaged in the process of bidding to stage a meeting and it is to the experience of the towns which participated in this process that attention is now turned.

Chapter 2
The Bidding Contest

The towns chosen to host one of the great annual parliaments of science were places which had come through often lengthy processes of scrutiny and selection in which their facilities and attractions were compared with those of rival venues. This chapter examines these processes, conducted with varying levels of procedural formality by the four knowledge associations, and investigates the strategies employed by applicant towns to attract one of their meetings. Aspiring meeting places had to offer not only the required practical facilities but also convince the hierarchy of the associations that they possessed a range of far more intangible qualities. Was there sufficient appetite for learning in the town and locality to sustain interest in a week-long meeting? Was there a local intelligentsia with the drive and enthusiasm to take on the burden of organisation? Would the most influential people in the town and neighbourhood throw their weight behind the meeting and lend their status to the event? Despite the rather impressionistic nature of these measures, for some towns they proved no obstacle to success. Places which had an established range of cultural institutions, a track record as a hub of social life and leisure, particularly for neighbouring gentry, and an ability to muster a deputation of influential supporters willing to petition for a meeting were likely to distinguish themselves above their rivals in the race for selection. It will be shown here that there emerged, during the half-century after 1831, a top tier of towns which were particularly successful in promoting themselves as meeting places.

The broader purpose of this chapter, however, is to suggest that these bidding contests for the annual parliaments of science should be viewed as part of a culture of inter-town competition and rivalry in urban Britain. Rivalry and comparison with other places was central to the process of urban identity formation in provincial towns. 'Competition and contrast', according to Joyce Ellis, 'lay at the heart of civic consciousness and identity' in eighteenth-century England.[1] Since the 'urban renaissance', when the cultural activities of the elite moved out of the country house and metropolis and into the clubs and societies, gardens and assembly rooms, provincial towns became 'an increasingly popular site for battles over status'.[2] The publication of guidebooks and other literature describing the appearance and character of towns, and peoples' own experience of places

---

[1] Joyce M. Ellis, 'For the Honour of the Town: comparison, competition and civic identity in eighteenth-century England', *Urban History*, 30/3 (2003): 325–37.

[2] J.J. Looney, 'Cultural Life in the Provinces. Leeds and York, 1720–1820', in A.L. Beier, David Cannadine and James M. Rosenheim (eds), *The First Modern Society. Essays in English History in Honour of Lawrence Stone* (Cambridge, 1989), p. 498.

based on increased travel and migration all helped to sharpen contemporary awareness of towns' status or urban rank. Leading towns could be distinguished by their wealthy appearance, their range of clubs and societies and their official functions as an assize town or market centre, serving the county or the wider region.[3] County towns in particular, with their inns, newspapers, trading floors and leisure venues, were 'inland entrepôts' serving not only their own townspeople but also the inhabitants of the surrounding country.[4] But, by the first half of the nineteenth century, industrial and demographic growth in some regions had given rise to burgeoning new communities whose size and wealth outstripped that of more established places. New industrial centres in the West Midlands, the North-West and Yorkshire were contributing to a major re-alignment of the urban map in Britain by 1841.[5] Thomas Carlyle's description of Leeds in 1847 as 'a huge smoky monster of a broadcloth town'[6] gives some indication of the impression that such places were making in the period. The towns which felt most under threat in the shadow of the new industrial metropolises were neighbouring historic county and market centres whose traditional position as the urban hub of their county or region was no longer guaranteed. Competition could be especially intense between neighbouring towns, where news of an improvement, honour or commercial fillip in one place could quickly generate fear of inferiority or of being 'left behind' in its near neighbours. Such sentiments often lay at the core of the battles over status, as in the Staffordshire Pottery towns in the later nineteenth century.[7]

Inter-urban rivalries manifested themselves in many different ways in the competitive urban climate of provincial Britain. Intensive dock building in the coastal towns of south Wales in the early to mid-nineteenth century saw rival ports compete with one another for trade; aggressive marketing and promotion, in many cases led by railway companies, was undertaken in the resort towns of the British coast in their competition to attract visitors. But the most important tool that towns utilised in their quest to compete was 'the pursuit of a civilising agenda'.[8] Iconic new town halls, environment-enhancing facilities like public parks and cultural facilities such as museums, libraries and theatres have all been identified as elements of the 'cultural capital' used by nineteenth-century towns

---

[3] Ellis, 'Regional and County Centres, 1700–1840', p. 674.

[4] Alan Everitt, 'Country, county and town: patterns of regional evolution in England', in Peter N. Borsay, (ed.), *The Eighteenth-Century Town. A Reader in English Urban History, 1688–1820* (London, 1990), pp. 83–115.

[5] John Langton, 'Urban growth and economic change, c.1688–1841', in Clark (ed.), *CUHB II*, p. 466.

[6] Carlyle Letters Online [CLO] 2007, http://carlyleletters.org, Thomas Carlyle to Jean Carlyle Aitken; 4 September 1847, accessed 6 October 2011.

[7] Jon Stobart, 'Identity, competition and place promotion in the Five Towns', *Urban History*, 30/2 (2003): 163–82.

[8] Stephen V. Ward, *Selling Places: The Marketing and Promotion of Towns and Cities, 1850–2000* (London, 1998), pp. 27–32.

to distinguish themselves from their rivals.[9] It is argued here that the hosting of parliaments of science, though they have not previously been considered in this context, were viewed in much the same way and were sought after by towns for the status-enhancing benefits they were believed to confer on their chosen locations.

For any Victorian town aspiring to host a parliament of science, the first stage in realising this ambition was to rally sufficient support from townspeople and urban leaders to issue a formal invitation to one of the country's leading peripatetic knowledge associations. Public meetings and town council debates were the usual settings where the pros and cons were aired and, in the discussions which took place in many towns about the merits of hosting a parliament of science, the rhetoric of comparison and the fear of inferiority loomed large. At a quarterly meeting of York City Council in 1843, for example, Alderman Hudson proposed that the Royal Agricultural Society of England be invited to the town to hold its show. Recently elected a member of RASE, Hudson had attended the show at Derby that year and reported to his fellow councillors that 'nothing could be more gratifying to him than the scene which he then witnessed'. While their own local agricultural society shows had been held in the barrack yard, 'he could not but feel the inferiority of the arrangements to the Royal Agricultural Society'.[10] At Swansea in 1851, when councillors deliberated the merits of bidding for the RASE show the following year, news that other Welsh towns were already ahead in issuing invitations had a direct bearing on the discussions. Dr Bird, who proposed the idea, pointed out that the town of Newport had already forwarded an invitation, 'and no doubt Swansea has a clear right to compete for the honour with any place'.[11] A sense of indignation at seeing other towns favoured with meetings, and an awareness of the benefits that such places had derived, was often detectable in the views of town council members who debated the subject. At a meeting of Nottingham town council in August 1863, Alderman Felkin noted that during its 32-year history, the British Association for the Advancement of Science had held its annual meeting in a number of places where 'the population and business are much smaller than in Nottingham and its vicinity'.[12] He viewed the BAAS meeting as a means to bring the best assets of the town to wider public attention, expressing 'no doubt that there were young men in this community who would stand comparison with those of any other' and adding his belief that BAAS members 'did not know what was the importance and extent of the town'.[13] In contrast, Newcastle-upon-Tyne was already receiving the British Association for the second time, and the Nottingham men reflected on how the BAAS visits had augmented the physical improvements of Newcastle,

---

9 Stobart, 'Identity, competition and place promotion', p. 166.

10 *York Herald and General Advertiser*, 12 August 1843.

11 *Cambrian*, 20 June 1851.

12 William Felkin was a Derbyshire-born lace manufacturer, evangelical Baptist and twice mayor of Nottingham in 1850 and 1851. See Roy A. Church, *Economic and Social Change in a Midland Town: Victorian Nottingham, 1815–1900* (London, 1966), pp. 326–9.

13 *Nottinghamshire Guardian*, 7 August 1863.

masterminded by the builder and property developer, Richard Grainger, and concluded that BAAS members returning there for their first visit since 1838 would see 'an agreeable connection of the progress in science and art, and in effort, which had made Newcastle what it now is'.[14]

As far as British Association meetings were concerned, there was a genuine anticipation of scientific and educational benefits and that these might be widely felt by residents. The Mayor of Leeds, addressing members of the town council in 1857, argued that 'There could be no doubt that such a meeting in Leeds would give an impetus to scientific pursuits in the town and neighbourhood, as in other towns the Association had visited'.[15] In Belfast, too, it was the vision of a generation of youngsters who might grow up to distinguish themselves by their scientific attainments that was invoked by one enthusiastic BAAS advocate who exhorted his townsmen to give their financial support to the meeting.[16] Elsewhere, it was the financial benefits for the local population which were emphasised. In Nottingham, Alderman Birkin suggested that: 'The visit of perhaps a thousand strangers in the town could not fail to benefit the place in a pecuniary way'.[17] RASE shows, too, were prized for their income-generating potential. The editors of the *Preston Guardian* put it very plainly in May 1868 when they concluded that, had Preston been selected to host the Royal Agricultural Society show the following year, 'a very handsome amount of money would have come into the town; and *that* is what we are all after'.[18] In Birmingham, when the merits of inviting the NAPSS for a second visit were being debated in 1867, the likely financial gains were set alongside other anticipated spin-offs. Harking back to the 1857 Social Science congress in the town, the mayor, Mr Avery, reminded councillors of the 'persons of highest eminence and attainment in science' who had attended, including Lord John Russell, Lord Stanley, Sir John Pakington, Professor Fawcett and Lord Brougham, and concluded that 'Both the material prosperity and the reputation of the town received substantial benefit from visits of this character'.[19]

There was not always unanimity in town council chambers and at public meetings when the merits of hosting a scientific meeting were being discussed. Voices of dissent were heard at Nottingham in 1865, when Alderman Birkin admitted that, 'Some gentlemen had asked him what advantage would the town get [*sic.*] by the British Association coming to Nottingham'.[20] In this case, Birkin managed to convince the doubters, but elsewhere attempts to rally support behind a meeting invitation fell flat. In 1876 Swansea surgeon W.H. Michael announced

---

14 Ibid.

15 *Leeds Mercury*, 13 August 1857.

16 Richard Davidson, esq., addressing a public meeting in the Corn Exchange, *Belfast Newsletter*, 16 April 1852.

17 *Nottinghamshire Guardian*, 11 August 1865.

18 *Preston Guardian*, 9 May 1868.

19 *Birmingham Daily Post*, 20 November 1867.

20 *Nottinghamshire Guardian*, 11 August 1865.

that he had received a communication from the Social Science Association stating its intention 'to hold their annual congress in Swansea, in September or October 1877'.[21] Michael, who had attended NAPSS congresses at Leeds in 1871 and at Plymouth in 1872, speaking on the subject of sanitation and sewerage on both occasions, was something of a lone enthusiast in Swansea. When a meeting was called by the mayor to discuss issuing a formal invitation to the Association, only six people attended, and despite an attempt by the editor of the town's newspaper to rally additional support,[22] the proposal got no further.

Such experiences illustrate that the enthusiasm of a few vocal individuals, coupled with a strong sense of competition with rival towns, was not enough to generate support for a meeting. Invitations had to reflect the strength of local cultural interests rooted in the clubs and societies of the district. Although most sizeable towns could boast a scientific or philosophical society of some description by the middle decades of the nineteenth century, these varied considerably in emphasis and specialism according to the character of each place. County towns drew on their wide occupational base and their established functions as social centres to develop as important hubs of provincial science in the eighteenth century, and sustained this role into the Victorian period, while the larger ports and industrial towns spawned a more utilitarian science culture.[23] Old regional capitals like Bristol, York, Norwich and Exeter developed broad-based institutions of science and culture in the eighteenth century which attracted participation from the established elites.[24] In the case of York, there was the added benefit of decades as the social capital of the north of England with an established season of concerts and assemblies frequented by the gentry of the region.[25] But by the late Georgian period, the historian of Britain's clubs and societies argued that there was a momentum shift in associational development away from these older centres and towards the 'rising industrial and commercial cities like Birmingham and Liverpool'.[26] The latter town, since the late eighteenth century, had established the Royal Institution of Liverpool, an Athenaeum, an Academy of Art and a Literary and Philosophical Society.[27] Manchester, meanwhile, had a Literary and

---

21 *Cambrian*, 11 August 1876.

22 Ibid.

23 Paul Elliott, 'Towards a geography of English scientific culture: provincial identity and literary and philosophical culture in the English county town, 1750–1850', *Urban History*, 32/3 (2005): 391–412.

24 See for example, Michael Neve, 'Science in a Commercial City: Bristol, 1820–1860', in Inkster and Morrell (eds), *Metropolis and Province*, pp. 186–97.

25 Looney, 'Cultural Life in the Provinces', p. 499.

26 Peter Clark, *British Clubs and Societies, 1580–1800. The Origins of an Associational World* (Oxford, 2001), p. 133.

27 Arline Wilson, '"The Florence of the North"? The Civic Culture of Liverpool in the Early Nineteenth Century', in Alan Kidd and David Nicholls (eds), *Gender, Civic Culture and Consumerism. Middle-Class Identity in Britain, 1800–1940* (Manchester, 1999), pp. 34–46.

Philosophical Society dating from 1781 and later, Britain's first society dedicated to the gathering of social statistics.[28] Here, and in the next tier of industrial towns like Sheffield, Bradford and Leeds, cultural and scientific societies were less the repositories of established wealth and status, and peopled more by the marginal or upwardly mobile in society.[29]

The nature of the knowledge landscape in a town was important from the very beginning of the bidding contest. It was often from the personnel of local cultural and scientific institutions that the initiative to bid for a meeting originated and it was their interests which influenced the type of congress they aspired to host. A few examples will serve to show how this operated in practice. By the 1830s Newcastle-upon-Tyne was home to a Literary and Philosophical Society, established in 1793, a Society of Antiquaries of Newcastle-upon-Tyne, established in 1813, and the Natural History Society of Northumberland, Durham and Newcastle-upon-Tyne, founded in 1829.[30] The Society of Antiquaries, in particular, with its membership of local gentry, clergy and lawyers,[31] helped make Newcastle an appealing venue for the Royal Archaeological Institute. It was their invitation, and the encouragement of their patron, the Duke of Northumberland, which 'gave a strong impulse' to the selection of Newcastle for the 1852 annual meeting of the RAI.[32] In contrast, the town had not spawned many leading social science enquirers among its intellectual elite. Its emerging brand of scientific knowledge was not typically 'applied' and the new social sciences gained few enthusiasts in the town when compared with other branches of enquiry such as astronomy.[33] Consequently, it was not until 1870 that the Social Science Association held its first annual meeting there, despite the fact that Newcastle was a popular venue on the annual meetings circuit by this date. NAPSS representatives visited it as early as 1862 to assess its potential as a meeting venue, but concluded only that, 'it is a good second-rate place, which we may wait for'.[34] With a population growth rate which was steady until the second half of the nineteenth century, it was perhaps not the archetypal industrial town that the NAPSS had come to favour, where

---

28 Thackray, 'Natural Knowledge in Cultural Context', pp. 672–709.

29 See for example, Ian Inkster, 'The Development of a Scientific Community in Sheffield, 1790–1850: a network of people and interests', in Ian Inkster, *Scientific Culture and Urbanisation in Industrial Britain* (Aldershot, 1997), pp. 102–17.

30 Orange, 'Rational Dissent', in Inkster and Morrell (eds), *Metropolis and Province*, p. 205.

31 George Jobey, 'The Society of Antiquaries of Newcastle-upon-Tyne', *Archaeologia Aeliana*, 5th series, 18 (1990): 208–9.

32 SAL, WAY MS 790, RAI Meeting at Newcastle 1852.

33 Frank Musgrove, 'Patterns of Intellectual Creativity in the North of England', *Northern History*, 33 (1997): 1–35.

34 UCL Special Collections, B MSS, 13,013, George W. Hastings to Lord Brougham, 1 September 1862.

statistical enquiry and the pressing social and economic questions of the day could be examined in situ.

For different reasons, throughout its entire lifespan, the NAPSS never held a meeting in Wales. There was ambivalence towards the social sciences in cultural circles in nineteenth-century Wales. Attempts to generate participation in social science debates had been made – notably with the introduction of a social science day in the Eisteddfod from 1862 – but small audiences led to the initiative being dropped.[35] As we saw earlier from the approach made to Swansea, there was evidence that the Association would have liked to have added a Welsh town to its list of meeting venues, but the idea failed to spark interest in a town where learned culture was focused around a scientific institution whose members favoured chemistry, geology and natural history over social investigation.[36] The British Association, in contrast, was regarded as a much more desirable prize and leading lights from the town's Royal Institution led a determined campaign in the 1840s to attract the BAAS to Swansea. Home-grown scientist William Robert Grove, nationally renowned for his pioneering work in developing the gas battery,[37] played a crucial role in persuading the town council to issue an invitation, and in representing the town's case to the BAAS.[38] Having people like Grove, who were not just enthusiastic townsmen, but active practitioners in a relevant field of knowledge, prepared to fly the flag on behalf of a town was a great boon.

In contrast, Bradford, off the BAAS radar until 1873, hosted only the third meeting of the NAPSS in 1859. Its failure to foster an enduring Literary and Philosophical Society, despite several attempts by local botanist Samuel Hailstone in the early nineteenth century, illustrated the extent to which this largely working-class town was divorced from the kind of science represented by the British Association.[39] Yet the paucity of traditional 'Lit and Phil' style science culture in the town which, as one historian has noted, struggled to find a secure foothold in the mushrooming urban textile centre of the early nineteenth century,[40] did not hinder the development of other types of scientific institution. The West

---

35 H.G. Williams, 'All Advance from Barbarism to Civilization is the Development of Social Science': Wales and the Promotion of Social Science', *Welsh History Review*, 20/4 (2001): 734.

36 A Swansea Statistical Society was active briefly in the 1830s. Louise Miskell, *Intelligent Town. An Urban History of Swansea* (Cardiff, 2006), p. 119.

37 Iwan Rhys Morus, 'Grove, Sir William Robert (1811–1896)', *Oxford Dictionary of National Biography*, Oxford, online edn, May 2005), accessed 12 Oct 2011.

38 For further details see Louise Miskell, 'The Making of a new "Welsh Metropolis": Science, Leisure and Industry in Early Nineteenth-Century Swansea', *History*, 88/1 (2003): 32–52.

39 Theodore Koditschek, *Class Formation and Urban Industrial Society, Bradford, 1750–1850* (Cambridge, 1990), p. 148.

40 Jack Morrell, 'Wissenschaft in Worstedopolis: Public Science in Bradford, 1800–1850', *British Journal of the History of Science*, 18 (1985): 1–23.

Riding Geological and Polytechnic Society, founded in 1837, was much more a product of the industrial and economic characteristics of the region and promoted the collection of statistical data on the coal and iron-working population.[41] In addition to this there was an enthusiastic social science contingent in Bradford's business community, led by local solicitor and secretary of the town's Chamber of Commerce, John Darlington. Darlington was known to the hierarchy of the NAPSS as, 'a man of considerable ability and information, having paid attention to social questions', who had 'joined our association at its first meeting and has worked for it ever since'.[42] Ultimately, a deep-rooted interest in the aims and activities of the association among a significant section of the leading inhabitants of a prospective host town and its region was a necessary precursor to any meeting invitation, let alone a successful bid. Bidding for a meeting, though it was a promotional undertaking, was about more than a simple desire to host an attention-grabbing event.

Like John Darlington in Bradford, other towns visited by the peripatetic associations for their annual meetings also had their influential knowledge ambassadors. John Towne Danson, the Liverpool entrepreneur and statistician, was a vital link between the national hierarchy of the NAPSS and a local Liverpool branch, which he helped to set up and run, as well as attending and presenting papers at several meetings of the parent organisation. He was an active member of the organising committee for the Liverpool meeting in 1858 and corresponded with George Hastings long after the 1858 meeting over political developments in social science policy.[43] Danson and Darlington exemplified the kind of men who peopled the cultural and scientific institutions of provincial Britain in the Victorian period. If not quite 'marginal' men, they were certainly socially up-and-coming and not part of the established elite of society in their respective regions. Men of their ilk drove the cultural life of many of Britain's growing ports, industrial and manufacturing centres, often producing a more innovative and novel brand of scientific enquiry than was evident in the more socially elevated institutions typical of older urban centres.[44]

The role of local knowledge enthusiasts, if it was important in persuading fellow townspeople of the merits of bidding for a meeting, became even more significant when the time came for the town's case to be put before the national association. As we saw in Chapter 1, where multiple invitations were received from aspiring host towns, as was often the case with the BAAS and RASE,

---

41 Jack B. Morrell, 'The Early Yorkshire Geological and Polytechnic Society: a reconsideration', *Annals of Science*, 45/2 (1988): 160–165.

42 UCL Special Collections, B MSS 19, 878, George W. Hastings to Lord Brougham, 21 February 1860.

43 See for example, NML MAL, John Towne Danson, D/D/III/2/62. Loose letters of George W. Hastings of the NAPSS to John T. Danson, 9 November 1859.

44 Ian Inkster, 'Introduction: aspects of the history of science and science culture in Britain, 1780–1850 and beyond', in Inkster and Morrell (eds), *Metropolis and Province*, p. 35.

competing delegations were invited to appear before the association's council to make their case for selection. The ability of the town to muster a powerful delegation, with the skills to convince a national association that it could host a successful meeting, was the crucial next step in the bidding contest. The composition of the delegation was all important. A high level of organisation and unity of purpose among the urban elite of the applicant town was required to bring together a group of supporters, preferably including representatives from local learned institutions, municipal bodies and gentry families. Moreover, these had to be people with sustained interest and commitment, prepared to repeat their work for the delegation over successive years if requests were initially unsuccessful. Details of the various deputations in attendance at the Cambridge BAAS meeting in 1862 provide a revealing insight into the powerful array of municipal and cultural talent that provincial towns were able to muster in support of their claims.[45] They show that the men (and in this period they were exclusively male) chosen to present the merits of their towns to the councils of the learned associations were typically people who combined prosperity and commercial success with political influence and also scientific or cultural distinctions. The Newcastle deputation that year included two prominent industrialists who had also distinguished themselves in national scientific circles. William G. Armstrong was head of an engineering business at Elswick, near Newcastle-upon-Tyne, a pioneer of hydraulic-powered machinery, armament maker and supporter of numerous local scientific societies.[46] Isaac Lowthian Bell was a chemical industry manager, fellow of the Royal Society and expert in the science of iron manufacture.[47] From Birmingham, the deputation included James T. Chance, the glass manufacturer and acknowledged pioneer in the development of effective lighthouse lenses.[48] Among the advocates for Bath were the town's mayor and William Tite M.P., president of the Royal Institute of British Architects and Fellow of the Society of Antiquaries.[49] The Nottingham contingent, meanwhile, included William Felkin, the lace and hosiery manufacturer, corporation member, erstwhile mayor, mechanics' institute promoter, statistician and historian.[50] The Royal Agricultural Society of England, likewise, made its decision on venues for forthcoming

45 Bodleian Library, Dep. BAAS 147, printed material relating to the Cambridge meeting, 1862.

46 Stafford M. Linsley, 'Armstrong, William George, Baron Armstrong (1810–1900)', *Oxford Dictionary of National Biography* (Oxford, online ed., 2006), accessed 13 Oct. 2011.

47 W.A. Campbell, 'Men of Science in nineteenth-century Newcastle', *Archaeologia Aeliana*, 5th series, 8 (1980): 60.

48 Charles Welch, 'Chance, Sir James Timmins, first baronet (1814–1902)', rev. A. McConnell, *Oxford Dictionary of National Biography* (Oxford, 2004), accessed 8 April 2009.

49 S.P. Parissien, 'Tite, Sir William (1798–1873)', *Oxford Dictionary of National Biography* (Oxford, 2004), accessed 8 April 2009.

50 Stanley D. Chapman, 'Felkin, William (1795–1874)', *Oxford Dictionary of National Biography* (Oxford, 2004, online ed., Jan. 2008), accessed 8 April 2009.

meetings after hearing the arguments put forward by visiting deputations made up of corporation representatives, members of local agricultural societies and other provincial associations. Here, too, representations made by figures from landed, municipal and industrial quarters, especially those with strong links to the major cultural institutions of the town and region, proved difficult to resist. In 1871, for instance, Cardiff mustered the support of the town's mayor, the county MP, the leading landowner and an enlightened industrialist to press its case to the RASE hierarchy, with the result that the annual show was held there in 1872.[51] The MP in question, Henry Hussey Vivian, was an industrialist engaged in non-ferrous metal manufacture, who also farmed at Gower, near Swansea, where he was known as a successful stock-breeder.[52] He was, in many ways, an ideal RASE delegation member: not only wealthy and influential, but also a man who embodied some of the interests and values of the Society.

As well as fielding politicians and industrialists who could demonstrate appropriate knowledge and expertise, towns also tried to impress the hierarchy of the learned associations by calling on the support of the gentry of the neighbourhood. A few big names and a sense of social status in local cultural ranks could be a significant help to the cause of an applicant town. From the very beginning, William Harcourt had appreciated the need for a firm endorsement by the gentry and social elites in order to secure the success of the inaugural BAAS meeting at York. Murchison noted that, by 'engaging his father [Archbishop of York] to act as a Patron, and by inducing Earl Fitzwilliam to be the President, he gave at once a *locus standi* and respectability to the project'.[53] In the same way, participation by the major landowners and noblemen of the district in a delegation could lend status and credibility to a town's application to host a parliament of science. Aristocratic support was particularly helpful to towns applying to host meetings of the Royal Agricultural Society. Along with Henry Hussey Vivian, Cardiff's deputation to RASE in 1871 included the Marquis of Bute, whose support for Cardiff's bid for the 1872 show on the grounds that it would be of 'benefit to his tenants', must have made a considerable impression on RASE council members who voted in favour of Cardiff over Cheltenham by 31 votes to six.[54] Lord Leigh's appearance on behalf of Warwick in 1858 was also instrumental in securing its choice for the 1859 RASE show, after losing

---

[51] MERL, SR RASE/B/I/7, Minutes of Council 1869–81 (rough copy), 3 May 1871. The deputation consisted of the town mayor, the Marquis of Bute, Mr Vivian MP and George T. Clark.

[52] W.R. Willliams 'Vivian, Henry Hussey, first Baron Swansea (1821–1894)', rev. Edmund Newell, *Oxford Dictionary of National Biography* (Oxford, online edn., 2006), accessed 13 Oct 2011.

[53] Geikie, *Life of Sir Roderick Murchison, volume 1*, p.185. *Locus standi*, noun, 'recognised position', *Oxford English Dictionary*, online version, accessed 15 April 2012.

[54] MERL, SR RASE/B/I/7 Minutes of Council, 1869–81, 3 May 1871.

out to Northampton 12 years earlier.[55] Similarly, in 1867 the Duke of Rutland's attendance as part of a Leicester delegation along with the town mayor convinced RASE that the town and county could 'pull together' and helped it to win out over rival applications from Derby and Nottingham.[56]

Towns which had close links with the neighbouring gentry thus stood the best chance of success in attracting a Royal Agricultural Society show. Such connections were strong in traditional county towns like Lincoln where the associational life of the urban elite and the neighbouring gentry might easily overlap at dinners, shooting parties and other social events,[57] a point which goes a long way to explaining the enduring popularity of these centres with RASE. But newer, industrial towns, too, were sometimes closely bound up with landed interests. Gentry ownership of land in and around provincial towns was often sufficient to give them a direct influence over urban development.[58] Those with mineral wealth on their estates became even more closely enmeshed with urban industrial economies as proprietors of collieries and ironworks employing large numbers of working-class town dwellers.[59] The inter-connectedness of a wide range of towns with their surrounding countryside enabled many of them to draw on a landed support when it came to making a case to host a parliament of science, and it was not just RASE, but the other associations too, which encouraged bidders to exploit this. Sir David Brewster,[60] replying on behalf of the BAAS to a request for advice on the submission of an invitation from Dundee, 'the seat of the great linen trade of the United Kingdom',[61] advised that,

> It would be very desirable that Sir John Ogilvie, your member, should go to the Oxford meeting to plead the cause of Dundee. If he should not be able to do this

55 Warwick triumphed over Reading and Oxford in 1858 and the links with the Leigh family were strengthened in the twentieth century when a site on their estate at Stoneleigh was adopted by RASE as its permanent show ground in 1960. See MERL, SR RASE/B/I/5 Minutes of Council 1853–58 (rough copy), 5 May 1858; Goddard, *Harvests of Change*, p. 197.

56 MERL, SR RASE/B/I/6 Minutes of Council, 1858–68 (rough copy), 1 May 1867.

57 Mark Rothery, 'The Shooting Party: the associational cultures of rural and urban elites in the late nineteenth and early twentieth centuries', in Richard W. Hoyle (ed.), *Our Hunting Fathers. Field Sports in England after 1850* (Lancaster, 2007), pp. 96–118.

58 See for example J. Ross Wordie, 'Introduction', in Christopher W. Chalklin and J. Ross Wordie (eds), *Town and Countryside. The English Landowners in the National Economy* (London, 1989), pp. 20–21.

59 See for example Richard Trainor, 'Peers on an industrial frontier: the earls of Dartmouth and of Dudley in the Black Country, c.1810–1914', in David Cannadine (ed.), *Patricians, Power and Politics in Nineteenth-Century Towns* (Leicester, 1982), pp. 70–132.

60 Edinburgh scientist and founder member of the British Association for the Advancement of Science.

61 Bodleian Library, Dep. BAAS 59, Patrick H. Thoms to Sir David Brewster, 20 June 1860.

> he might write to Sir Roderick Murchison or any influential member with whom he may be acquainted. As Perth is interested in the matter, Mr A. Kinnaird might also exert his influence and I have no doubt that Earl Camperdown might also give his concurrence and aid in having a meeting at Dundee.[62]

Gentry involvement in annual meetings was not confined to the bidding stage. The associations were further reassured by evidence of their intention to participate in the meeting, by providing accommodation for distinguished guests, or opening up their estates to visitors during the course of the congress. The Duke of Northumberland fulfilled all of these functions during the visit of the NAPSS to Newcastle in 1870, accepting the presidency, delivering the opening address to delegates and hosting a visit to Alnwick Castle.[63] His participation epitomised the benefit which Newcastle derived from gentry involvement in its cultural life. Renowned for its large and hospitable gentry presence, Newcastle's reputation as a viable meeting centre was enhanced by their active involvement, and by the fact that the industrial and commercial elites of the town were perceived to be well integrated into their social circles.[64] But there was a reciprocal advantage to the landowner too. Seen as part of a broader repertoire of activities which included traditional acts of hospitality and festivities linked with their local benefactions,[65] gentry figures could cement their status and authority in rapidly growing urban centres by associating themselves with annual meetings in this way. However, not all regional metropolises could call on this level of gentry participation. Roderick Murchison wrote to William Harcourt in the Autumn after the 1838 Newcastle meeting and in the early phases of planning the following year's BAAS gathering at Birmingham, and observed that in Newcastle,

> … we found a sort of aristocratic bourgeoisie of bankers, glass makers etc who were connected with the best families in the county, rich and influential. Indeed, many of the Northumbrians of the largest landed estates owe their origin to the town. Here there is nothing of the sort, for Boulton and Watt, the Gog and Magog of the new era, have cut all connection with the place and will take no

---

[62] Bodleian Library, Dep. BAAS 59, Sir David Brewster to Patrick H. Thoms, n.d. Arthur Kinnaird was the tenth Lord Kinnaird of Inchture and second Baron Kinnaird of Rossie. He was MP for Perth from 1852–78. Earl Camperdown was Adam Haldane Duncan, grandson of Viscount Adam Duncan of Camperdown. See Frank Prochaska, 'Kinnaird, Arthur Fitzgerald (1814–87)'; and P.K. Crimmin, 'Duncan, Adam (1731–1804)', *Oxford Dictionary of National Biography* (Oxford, 2004), accessed 24 February 2012.

[63] *The Times*, 10 August 1870, 17 September 1870, 23 September 1870.

[64] Ellis, 'Regional and County Centres, 1700–1840', p. 691.

[65] See for example, Rick Trainor, 'Peers on an Industrial Frontier: the earls of Dartmouth and of Dudley in the Black Country, c.1810–1914', in David Cannadine (ed.), *Patricians, Power and Politics in Nineteenth-Century Towns* (Leicester, 1982), p. 106.

> part in the doings; so that the best people in the place are decidedly inferior to the Newcastle auxiliaries.[66]

With an established tradition as a social centre with a race week and assembly rooms hosting a range of events since 1776, Newcastle had already distinguished itself as 'the preferred venue for the annual round of socially prestigious occasions'.[67] Measured against these standards, the growing manufacturing and industrial centres, without the same leisure infrastructures to facilitate a bringing together of urban elite and neighbourhood gentry, found themselves at a disadvantage.

The status and effectiveness of the delegation chosen to present a town's bid for a parliament of science was just one factor in its likely success. Of equal or greater significance was the ability of the town to demonstrate that it had the practical facilities and capacity to host a meeting. The availability of suitable rooms or open spaces in which to hold the various debates, discussions, lectures and exhibitions that made up the annual meeting was a central consideration, making the provision of accommodation crucial for any aspiring meeting place. The requirement of housing both large general addresses and sectional meetings in a single building, or at least in proximate ones, was a key consideration for associations choosing their meeting venues.

Towns wishing to host meetings of the BAAS or NAPSS, which had multiple sections meeting simultaneously in different committees, needed to provide 'twelve or fourteen commodious rooms',[68] including a hall large enough to accommodate mass lectures, evening receptions and dinners. The trouble was that in the early years of BAAS in particular, few provincial towns and cities possessed such accommodation. It was no coincidence that four of the first five British Association meetings were held in Oxford, Cambridge, Edinburgh and Dublin where it was the universities which provided the meeting rooms and social spaces for delegates. Edinburgh University Library's Great Hall, Dublin's Trinity College and the Sheldonian Theatre and Clarendon Buildings at Oxford University all provided venues which few other provincial towns of the period could match. The following year, when Cambridge was chosen for the annual show of the Royal Agricultural Society of England, a letter was sent to the Master of Trinity College to request, 'the use of your Hall for a dinner of the RASE … no other room in Cambridge is sufficiently large for this purpose'.[69]

A university location for an annual meeting had the added benefit of conferring intellectual status on the association. For the BAAS, going to Oxford for its second

---

66 Roderick Murchison to Williams Harcourt, 8 November 1838, quoted in Morrell and Thackray (eds), *Gentlemen of Science. Early Correspondence*, pp. 293–4.

67 Helen Berry, 'Creating Polite Space: the organization and social function of the Newcastle Assembly Rooms', in Helen Berry and Jeremy Gregory (eds), *Creating and Consuming Culture in North-East England, 1660–1830* (Aldershot, 2004), p. 140.

68 *Leeds Mercury*, 13 August 1857.

69 MERL, SR RASE, B/XXXII/I, Letterbook, 1838–44, 11 June 1840.

Figure 2.1 The Geology Section in the Grammar School at the 1849 British Association meeting in Birmingham, *Illustrated London News*, 22 September 1849 (author's collection)

meeting helped to secure the future of the Association. Roderick Murchison later wrote that, 'It was that second meeting which consolidated us, and enabled us to take up a proper position. Then it was that, seeing the thing was going to succeed, the men of science of the metropolis and those of the universities joined us'.[70] RASE, too, saw the value of locating its earliest annual shows in university towns from the point of view of both intellectual status and practical facilities. RASE councillors wrote to the university vice chancellor after their 1838 meeting in Oxford, acknowledging that, 'The support of such high authority in the ancient university of Oxford cannot fail to be productive of beneficial effects in furthering

70 Geikie, *Life of Sir Roderick I. Murchison*, vol. 1, p. 188.

the objects of the society'.[71] But by taking advantage of the facilities at these 'well endowed, oligarchically self-governed' institutions,[72] the associations were not really embracing the urban location, and the role of the town as a scientific venue was not clear cut. When the BAAS visited Oxford in 1832, Alexander Croke, president of the Ashmolean Society, declared at the opening presidential address that it was 'a proud day for the University of Oxford', a comment which seemed to cast the university rather than the town in the role of official host.[73] Delegates attending a meeting accommodated in university buildings could be insulated from the wider surroundings of the town, as was apparent to one newspaper reporter attending the British Association meeting in Belfast in 1874. He observed that the Queen's College buildings, in a 'fashionable suburb of Belfast', were 'just clear of the noise of the factories below'. With their spacious grounds they provided ample recreational space for delegates between sessions so that 'outside this collegiate radius, the association does not require to go unless for public addresses'.[74] A similar claim might have been made by the University of Cambridge in 1862 when the meeting of the British Association was accommodated in the lecture rooms and halls of four different colleges.[75] Later in the century, the buildings belonging to the newly developing civic universities provided attractive, self-contained meeting spaces for municipal leaders to present to possible visiting bodies. The Lord Provost of Glasgow, who formed part of a powerful delegation at the Bradford meeting in 1873 to pitch for a British Association visit the following year, suggested that the new University would be complete in time for the meeting and that 'the whole building would be placed at the disposal' of the Association for the duration of the week.[76]

Beyond the rarefied surroundings of the universities, it was to the assembly rooms and leisure venues of the great tourist centres that the associations often ventured. The specialist resort and leisure towns of the Georgian era were, in many cases, well equipped with rooms and halls which the learned associations could utilise. The period from the late seventeenth to the mid-eighteenth century had spawned a new level of visitor-oriented facilities in some towns and, although the emphasis was on entertainment, with assemblies, theatres, races and concerts,[77] the by-product was an infrastructure which could be adapted for other types of gathering. Brighton, for instance, met with the approval of the Social Science

71 MERL, SR RASE, B/XXXII/I, Letterbook 1838–44, 2 August 1839.

72 David R. Jones, *The Origins of the Civic Universities. Manchester, Leeds and Liverpool* (London, 1988), p. 12.

73 Bodleian Library, Dep. BAAS 130, Notebook of John Phillips, 1832, p. 9.

74 *The Standard*, 20 August 1874.

75 Bodleian Library, Dep. BAAS 147, printed material relating to the Cambridge meeting, 1862.

76 *Leeds Mercury*, 23 September 1873.

77 Peter N. Borsay, 'The English Urban Renaissance: the development of provincial urban culture, 1680–1760', *Social History*, 5 (1977): 581–693.

Association in 1875 when it was found to be 'a great advantage to the members to have all the meetings held under one roof as they can be in the Royal Pavilion. The general address will be delivered under the dome and the sections will meet in adjacent rooms'.[78] The appropriation of such buildings for serious intellectual pursuits was potentially problematic for the learned association seeking intellectual credibility. Resort towns were noticeably scarce as meeting venues of the Archaeological Institute, which was particularly concerned with its reputation for scholarly work. Places with abundant tourist attractions may have appeared risky to an Association anxious to prevent its annual meetings from being over-run by 'too many people of whom too few were really antiquaries'.[79] It did not visit Bath until 1858, despite the town's appeal as a magnetic centre of social life and its ability to turn what were usually tourist facilities, notably its pump room and a theatre capable of accommodating some 1,700 people, to the use of visiting scientists and meeting delegates.[80] Such diverse preferences as those exhibited by the social scientists and the archaeologists illustrated the very different sets of priorities governing the relationships between the four associations and their host venues. There was no single meeting place model which had universal appeal.

If university towns and resorts tended to be well blessed with buildings large enough to accommodate learned meetings, the same was not necessarily true of the growing commercial and manufacturing centres. Those that were quick off the mark in erecting public halls reaped early rewards. Bradford's St George's Hall, built between 1851 and 1853, was 'a magnificent building, capable of seating in comfort 3,700 persons',[81] funded with £16,000 raised by public subscription.[82] As a manufacturing town with a capacious new venue it was thus high on the list of desirable destinations for the NAPSS which held only its third annual meeting there in 1859. Liverpool was also attractive to the NAPSS in part because it possessed 'a large hall for public meetings, with accommodation for five departments, if possible, under one roof'. George Hastings of the Social Science Association told the local organising committee at Liverpool that, 'If we can obtain St George's Hall, which I believe is one of the finest buildings in the country, we shall be most splendidly provided for'.[83]

---

78 *The Times*, 5 October 1875.

79 Joan Evans, 'The Royal Archaeological Institute: A Retrospect', *Archaeological Journal*, 106 (1951 for 1949): 1–11.

80 Penelope Corfield, 'Georgian Bath – the Magical Meeting Place', *History Today*, 40/11 (1990): 26–33. See for example the account of the British Association meeting at Bath in the *Leeds Mercury*, 14 September 1864.

81 *Daily News*, 12 October 1859. Note that Koditschek puts the capacity of the hall at 3,100. See Koditschek, *Class Formation*, p. 577.

82 Koditschek, *Class Formation*, p. 577.

83 NML MAL, John Towne Danson, D/D/III/15/3, Printed pamphlet: Proceedings of the meeting of the General Committee of the NAPSS on 5 May 1858.

Some of the most rapidly growing industrial centres of provincial Britain in the middle decades of the nineteenth century, however, found themselves wanting in this regard. Manchester, although it had support from influential BAAS members including Charles Babbage, William Whewell and Charles Daubeny for a meeting as early as 1833, lacked the type of accommodation required for a meeting. William Henry, the Manchester chemist and Literary and Philosophical Society member, was an early advocate of hosting an annual meeting of the BAAS but was forced to admit that, 'at present we have not in the town anything in the way of a suite of rooms, adapted for the various committees and public meetings of such a body'.[84] Babbage later conceded that, 'I was one of those, who at the first moment thought we should adjourn to some large manufacturing town; but I am now convinced that there would be fewer difficulties in our way if that were for a time postponed'.[85] It was another decade before the BAAS carried out its intention of holding a meeting in Manchester, and then it was still found 'impossible to accommodate the sections under one roof'. Instead the business was dispersed across a range of venues from the Royal Institution and the Athenaeum to the Friends Meeting House to the Assembly Rooms and Mechanics' Institution.[86] The environment of the town was altered by the time of the first meeting of the NAPSS there in 1866. Among the architectural improvements of the 1850s and 1860s were the re-development of the Piccadilly district and the appearance of grand law courts, a prison and asylum.[87] It was 'the beautiful New Assize Courts' which was adapted for use by the Social Scientists in 1866. Its size meant that, 'Not only will all the business of the Congress be done there, but on the evenings of the 4th and 8th of October the courts are to be the scene of two brilliant soirees'.[88]

Ambitious planning and improvement projects, especially those which incorporated capacious new buildings, made an impression on the national associations and their members. When the BAAS visited Newcastle-upon-Tyne in 1838, delegates were immediately struck by the urban grandeur of the town re-fashioned by Richard Grainger, well before any similar 'processes of aggrandisement' had been completed in other provincial centres.[89] The quality of accommodation on offer to the different sections of the meeting made a profound impression on John Phillips who wrote to William Harcourt on the eve of the annual gathering that:

---

84 Frank Greenaway, 'Henry, William (1774–1836), *Oxford Dictionary of National Biography* (2004, online ed., 2007), accessed 24 February 2012. Quoted in, Morrell and Thackray, *Gentlemen of Science. Early Years*, p. 167.

85 Bodleian, Dep. BAAS 130, Notebook of John Phillips, 1832, p. 19.

86 *Manchester Times and Gazette*, 25 June 1842.

87 Simon Gunn, 'The middle class, modernity and the provincial city: Manchester c.1840–1880', in Kidd and Nicholls (eds), *Gender, Civic Culture and Consumerism*, pp. 116–7.

88 MCL, BR F306 N1, extract from *The Builder*, 8 September 1866.

89 Ian Ayris, *A City of Palaces. Richard Grainger and the Making of Newcastle-upon-Tyne* (Newcastle, 1997), p. 15.

> Newcastle, which was one of the blackest and filthiest old carbonaceous towns in the land, has been utterly transformed so as to rival in many of its parts the best parts of Edinburgh: or rather, to say merely the truth, to *surpass* them. For the style of the buildings is admirable, and their masses really noble ... The evening business [is] in a room (quite splendid) built by the man who has transformed Newcastle (Grainger), holding 2700 seats and 300 standings.[90]

Phillips' comments indicate that improvements to the urban fabric of a town could transform its status as a potential meeting venue in the eyes of national associations. In the case of Newcastle, further meetings of the Royal Archaeological Institute in 1852 and 1884, the Social Science Association in 1870 and the British Association again in 1863 confirmed its broad appeal as a venue capable of accommodating a large congress.

Figure 2.2 Lord Brougham addresses the 1865 Social Science Congress in Sheffield at the Alexandra Music Hall, *Illustrated London News*, 14 October 1865 (author's collection)

[90] John Phillips to William Harcourt, 10 August 1838, quoted in Morrell and Thackray (eds), *Gentlemen of Science. Early Correspondence*, pp. 272–3.

Other industrial towns felt themselves disadvantaged because they lacked such spacious public accommodation. In 1863, one Sheffield Alderman, taking part in a council debate on the merits of a new town hall scheme, pointed out that 'we should, if we had such a building, be able to invite the Social Science Association and the British Association to hold their meetings in this town'.[91] Sheffield's new town hall was not completed until 1897,[92] but in the meantime the accommodation deficit was circumvented by developments to existing buildings. The Social Science Association's congress was staged in the town in 1865, utilising the old Adelphi Theatre premises, which was re-developed and re-branded as the Alexandra Music Hall in time for the meeting.[93] In contrast, the neighbouring textile town of Bradford succeeded in hosting the British Association for the first time just days after the formal opening of its new Exchange Hall in September 1873, which was used as the main visitor reception for the duration of the meeting. The timing of the meeting to coincide with the opening of the new hall was a public relations *coup* for the town. Visiting members of the press hailed the new Exchange as 'a first-class reception room', and concluded that 'Those who have not seen Bradford during the last few years will be astonished at the progress it has made ... there have sprung up a number of really handsome buildings'.[94] Similarly in Ipswich, where a new Corn Exchange was opened in 1850, townspeople saw the potential of their new building not just as a venue for the commercial transactions of the farming and merchant community, but also for public meetings, concerts and other gatherings. With almost 5,000 square feet of meeting space, capable of seating 1,000 or accommodating a standing audience of more than double that number,[95] the potential of the new building was quickly seized upon. A deputation from the town attended the Edinburgh meeting of the BAAS that summer and succeeded in convincing them to select Ipswich as the location for the 1851 annual meeting. One member of the triumphant delegation shrewdly noted that, 'But for the erection of this hall, or one of similar dimensions, it would have been impossible for Ipswich to have received the visit of the British Association'.[96]

In stark contrast to the success of Ipswich, seven invitations were issued to the BAAS from the town of Leeds between 1850 and 1857 before it was finally chosen to host the annual meeting in 1858.[97] Leeds' failure to attract a large meeting

---

91 Alderman Holland addressing a town council meeting, *Sheffield and Rotherham Independent*, 15 October 1863.

92 Colin Cunningham, *Victorian and Edwardian Town Halls* (London, 1981), p. 286.

93 *Morning Post*, 38 September 1865.

94 *Daily News*, 17 September 1873.

95 *Ipswich Journal*, 7 September 1850.

96 These comments were made by George Ransome during a public dinner at the new Corn Exchange. See *Ipswich Journal*, 14 September 1850.

97 Bodleian Library, Dep. BAAS 147, Conspectus of Invitations received and Meetings Held from 1848–62. Drawn up by John Phillips, Assistant General Secretary.

was keenly felt by the late 1850s when the mayor admitted that 'Leeds had not taken that position amongst the large towns of the country which she ought to have done',[98] but it was felt that the impending completion of a new town hall would change its fortunes. The construction, fitting out and opening of new public buildings here as elsewhere was carried out with visitations from distinguished bodies like the BAAS firmly in mind. At a meeting of Leeds town council in 1857, where it was decided once more to issue an invitation to the British Association, assurances were given that the town hall, under construction since 1853, would be finished in time for the meeting and it was agreed that the arrival of the BAAS would represent 'a fitting inauguration' for the new building.[99] The case of Leeds was symbolic of the problems facing new manufacturing and industrial towns with aspirations to establish themselves as centres on the conference circuit. It came close to hosting the Royal Agricultural Society of England when a deputation from the town pressed its advantages to the council in 1847 as a great centre of population, with manufacturers who would open up their works for inspection, well connected by railways and also in close proximity to open moorland. There were some RASE council members who were convinced by the claims. Pusey felt that the flax mills such as Marshall's would present an attraction to show-goers, and pointed out that the society had not yet tried a manufacturing town. It was also thought that implement-makers and farmers might benefit from mingling with manufacturers, but, ultimately, York's more traditional attractions as a cathedral and assize town with a music festival and with stronger agricultural connections won the day.[100]

The contest between York and Leeds for the RASE show in 1847 was an early indication of the growing regional rivalry between the two centres and it was surely no coincidence that, over the next decade, Leeds' municipal leaders fought a determined campaign to gain functions and status which they saw as befitting a town of its size and population. The prolonged campaign to host the BAAS coupled with the associated efforts to build a new town hall were part of a broader drive by urban leaders to establish Leeds as the principal town in the West Riding. Simultaneously they pressed for the re-location of the assize court from its traditional place in York, to the more populous part of the county. There was no reason, the editors of the *Leeds Mercury* opined in February 1857 in one of their many articles on the subject, why assize courts in populous counties had to be held in the official county town:

> The business of the West Riding ought as soon as practicable to be transferred to Leeds, as combining the advantages of being the most populous town, of having

---

[98] *Leeds Mercury*, 13 August 1857.

[99] *Leeds Mercury*, 13 August 1857.

[100] Details of this debate can be found in MERL, SR RASE/B/I/3, Minutes of Council 1844–48 (rough copy), 5 May 1847.

> the greatest railway facilities, and of possessing, in our noble and spacious Town Hall (when completed), a place worthy of the administration of justice.[101]

In 1857 and again in 1858, Leeds town councillors applied for removal of the assizes, but their efforts to become a legal and administrative centre for the West Riding were resisted not only by York, but by neighbouring Wakefield which harboured similar ambitions.[102] The Leeds case illustrates the extent to which the annual meetings circuit was bound up with patterns of inter-urban rivalry in this era.

The battle between York and Leeds was typical of the contests fought over the hosting of Royal Agricultural Society of England shows. Unlike the other associations in this study, RASE's method of visiting a different district each year on a rotational basis[103] meant that bidding competitions were always confined to towns located in the same region, often pitting new industrial and commercial centres against more traditional county towns. This had the effect of accentuating rivalries. In 1843, for example, when towns from the southern counties of England were invited to compete for the right to host the show the following year, a fierce battle ensued between Maidstone, Southampton and Winchester. Delegates from each town were invited to make their case at the society's council meeting, and their applications demonstrated the lengths to which they were prepared to go to be judged the best venue. Maidstone, which the society noted had the benefits of being 'a very large fine town ... of 18,000 inhabitants, 35 miles from London', declared itself 'willing to do anything' and 'quite ready to guarantee £500' towards the cost of hosting the meeting. The Society, however, worried about access, noting that, at the time, the town was ten miles away from a railway. Southampton's mayor raised the stakes, guaranteeing a fund of £1,200 towards the cost of the meeting and offering the society gratuitous use of any land and accommodation necessary for its needs. The council liked the fact that lodging accommodation seemed to be abundant in Southampton and that within a 20-mile radius there were some 21 market towns from where visitors might be drawn to attend. The Winchester delegates, meanwhile, promised to put the use of its college dining hall at the disposal of the Society. Its communication links were good, with its proposed show yard site lying within 300 yards of a railway station, and the council was convinced of a general 'feeling of goodwill' towards the Society.[104] In the end the decision was put to a vote and Southampton was selected. RASE's secretary wrote to the Winchester delegation with the news, but

---

101 *Leeds Mercury*, 3 February 1857.

102 A. E. Peacock, 'The creation of the West Riding Court of Assize', *Northern History*, 22 (1987): 119–37.

103 In its monthly council meeting in December 1841, for example, the next four districts to be visited were mapped out as Derbyshire in 1843, Middlesex in 1844, North Wales in 1845 and Northumberland in 1846, although these areas were not strictly adhered to. MERL, SR RASE B/I/1, Minutes of Council 1840–42 (rough copy), December 1841.

104 MERL, SR RASE B/I/2, Minutes of Council (rough copy), 3 May 1843.

probably failed to offer much comfort by pointing out that, 'in having selected Southampton as the place of their meeting they have fortunately chosen a town within the immediate sphere of communication with the city of Winchester, and hope accordingly for the cordial co-operation of their friends in that district'.[105]

Contests were even more keenly fought in regions where there was no dominant urban centre. The East Midlands towns of Derby, Leicester and Nottingham were sufficiently different and distant from one another 'to prevent any one of them from becoming dominant, despite the strong rivalry which developed between them'.[106] It made for a closely contested competition when the mayors and other representatives of all three competed head-to-head in 1842 and again in 1867 to host RASE. In 1842 Derby was judged preferable to Leicester because of its superior rail links. Early and effective railway development in the town marked it out as an important nodal point in the expanding transport system.[107] Nottingham's application, on this occasion, was not even put to the vote.[108] Twenty-five years later, a special committee of Nottingham town council was formed with the aim of planning an approach to the Society,[109] and the Mayor, Mr Thackray, pledged £2,000 to fund the event. Despite the financial promises, the Nottingham bid made little impact on RASE's governing council, attracting only two votes to Leicester's 26.[110] It had fielded a deputation of corporation officials and members of parliament, but had failed to demonstrate a united front of urban and county agricultural interests. One disappointed townsman urged the landowners and farmers of the county to re-form a Nottinghamshire Agricultural Society, and likewise exhorted townspeople to recognise the ways in which the 'practical knowledge of our farmers might be beneficially used' in important urban debates over health, water supply and sewage.[111] In a town dominated by hosiery and lace manufacture, the fusion of landed and industrial interests was much less pronounced than in places where the mineral and property wealth of gentry estates underpinned the urban economy. But where this hampered Nottingham's ability to impress the Royal Agricultural Society, it had more success with the BAAS. Considerable municipal resources were devoted to a four-year bidding campaign which saw successive town mayors appear in person to present invitations at BAAS meetings in Cambridge, Newcastle, Bath and Birmingham. It was a commitment which neither Leicester nor Derby undertook in the same period and the result was that Nottingham hosted the BAAS meeting in 1866. Its credentials as a congressional centre to rival that of its East Midlands neighbours were thus secured.

---

105 MERL, SR RASE B/XXX11/1, Letterbook, 1838–44, 11 May 1843.

106 Alan Dyer, 'Midlands', in Clark (ed.), *CUHB II*, p. 103.

107 John R. Kellett, *The Impact of Railways on Victorian Cities* (London, 1969), p. 3.

108 MERL, SR RASE B/I/1 Minutes of Council (rough copy), 4 May 1842.

109 *Records of the Borough of Nottingham. Being a series of extracts from the archives of the Corporation of Nottingham, vol. IX: 1836–1900* (Nottingham, 1956), p. 173.

110 MERL, SR RASE/B/I/6, Minutes of Council (rough copy) 1 May 1867.

111 Letter from J.C. Nixon to the editor, *Nottinghamshire Guardian*, 3 May 1867.

Increasing participation by mayors and other town officials in the selection processes had repercussions for competing venues. The involvement of local government leaders and other dignitaries put the civic reputation of the town on the line and made the annual contests a much more public affair. Municipal involvement might increase a town's chances of success, but it also multiplied the consequences of failure. No mayor wanted to suffer the indignity of losing out to a rival, but as the learned associations announced their decisions each year, unsuccessful towns were inevitably exposed and experienced a sense of wounded pride. Manchester for example, felt unjustly served by the British Association.

As Table 2.1 shows, its second BAAS meeting in 1861 came after the issuing of seven consecutive invitations. Notwithstanding the problems it had experienced in earlier decades with furnishing sufficient rooms to host a congress, there was a strong feeling that the town was being unfairly overlooked and that its place in the 'queue' of host venues had been usurped by less deserving places. William Fairbairn, an engineer and president of the Manchester Literary and Philosophical Society,[112] wrote to John Phillips in 1859 after gathering expressions of support from cultural and civic institutions in the town:

> I have seen the mayor this morning on the subject of the association holding its meeting in 1861 in this city and I enclose you the resolution of the Council of Wednesday last. I have also the resolutions of the Literary and Philosophical Society, the Mechanics Institution, the Geological and Natural History societies, from each of which there will be a deputation to invite the association for 1861 and I hope there will be no attempt – as at Dublin – to cut us out of our regular turn. I can assure you I have had some trouble to reconcile parties who were offended at the Dublin proceedings and in hopes we may be better treated next year at Oxford.[113]

In the event, Manchester got its meeting in 1861 and Fairbairn was nominated president, but the communication to Phillips during the course of the campaign revealed the extent to which success in securing annual meetings had come to be viewed as a measure of urban esteem. For large towns and cities especially, where bidding for a meeting was part of a wider agenda to advance their urban status, the prospect of being judged unsuitable, or of losing out to a 'lesser' place, could be a very public set-back, making the process of competing a risky business.

The same sentiments were evident in the aftermath of RASE contests. In 1868 for example, RASE invited bids from towns in its new 'northwestern district', comprising Lancashire and the West Riding, to host its show in 1869.[114] Lancaster,

---

112 See James Burnley, 'Fairbairn, Sir William, first baronet (1789–1874)', rev. R. Brown, *Oxford Dictionary of National Biography* (Oxford, 2004; online ed. May 2008).

113 Bodleian Library, Dep. BAAS 59, William Fairbairn to John Phillips, 8 September 1859.

114 MERL, SR RASE/B/I/6 Minutes of Council (rough copy), 4 December 1867.

Table 2.1 British Association for the Advancement of Science: invitations received and meetings held, 1848–1862

Town	**1848**	**1849**	**1850**	**1851**	**1852**	**1853**	**1854**	**1855**	**1856**	**1857**	**1858**	**1859**	**1860**	**1861**	**1862**
Swansea	Meeting														
Birmingham	*	Meeting													
Edinburgh		*	Meeting												
Ipswich	*	*	*	Meeting											
Belfast		*	*	*	Meeting										
Hull		*	*	*	*	Meeting									
Liverpool			*		*	*	Meeting								
Glasgow					*		*	Meeting							
Cheltenham							*	*	Meeting						
Dublin						*	*	*	*	Meeting					
Leeds			*	*	*	*		*	*	*	Meeting				
Aberdeen										*	*	Meeting			
Oxford										*	*	*	Meeting		
Manchester		*					*	*	*	*	*	*	*	Meeting	
Cambridge											*	*	*	*	Meeting

Source: Bodleian Library, Dep. BAAS 147, from information compiled by John Phillips, Assistant General Secretary. (*indicates invitation received)

Liverpool, Manchester and Preston all expressed interest in hosting the Society and the ensuing contest revealed much about the shifting urban order in the north west. An inspection committee visited the four towns to assess their suitability, but a month later, only the last three were invited to send delegations to RASE's council meeting for the final reckoning: Lancaster was left out. Preston's supporters showed little sympathy as, 'the old county town was soon thrown overboard'.[115] They based their claim to be the best choice on geographical centrality: 'Preston is really the best place in Lancashire for a great agricultural show. Its position – its centrality, its contiguity to excellent farming lands, its easiness of access from all parts, render it by far the most fitting place for holding a show like that named'.[116] But ultimately it was Manchester, backed by a town council resolution, a subscription of £8,000 and claiming a sphere of influence extending into neighbouring Cheshire, which won the approval of the Society, overhauling Preston by 25 votes to 18.[117] It was a decision which RASE members later admitted had 'offended everybody'.[118] The selection of Manchester by RASE for its 1869 show demonstrated the lure of the larger manufacturing and commercial centres. Although by the late 1860s Lancaster had passed its prime as an administrative and commercial centre for Lancashire,[119] it was no less painful an experience to have a kind of official recognition of this eclipse by its industrial neighbours confirmed by RASE at such an early point in the contest. The blow was softened somewhat by the fact that two months after the RASE decision in July 1868 Lancaster hosted the annual gathering of the Royal Archaeological Institute. It was the perfect opportunity for a show of defiance. The local press lost no time in reasserting Lancaster's importance, though presumably no ironic reference was intended to the recent snub by the Agricultural Society when it reported that, 'The county town has been chosen as the centre of a field of operations from which the [archaeological] society expect to glean a good harvest'.[120] Lancaster town council, meanwhile, in its welcome address to the Archaeological Institute, described their town as 'the capital of the county palatine'.[121] The county-town setting continued to suit the different scale and focus of the archaeological gatherings and it was a reflection of the different priorities of the learned associations in seeking urban meeting locations that Lancaster could almost simultaneously find favour with one while being rejected by another.

The difference between success and failure for applicant towns seeking to host a national association often came down to the issue of accessibility. The prime considerations for RASE in choosing meeting locations were the availability of

---

115 *Preston Guardian*, 9 May 1868.

116 Ibid.

117 MERL, SR RASE/B/I/7, Minutes of Council (rough copy), 6 May 1868.

118 MERL, SR RASE/B/I/7 Minutes of Council (rough copy), 7 December 1870.

119 John K. Walton, 'North', in Clark (ed.), *CUHB II*, pp. 118–9.

120 *Lancaster Gazette and General Advertiser*, 27 June 1868.

121 *Lancaster Gazette and General Advertiser*, 1 August 1868.

land for a showground and pavilion and the proximity of a railway line. [122] With farm implements and livestock to move to the location of its annual show, it is easy to see why RASE placed so much emphasis on good railway communication and, consequently, it was the ease of accessibility by rail, in particular, which saw most bids stand or fall. Such was the case at Worcester, which was rejected as a meeting venue by RASE in 1852, but re-applied successfully a decade later to host the 1863 meeting, by which time it had 'become the centre of a system of railways, radiating to all parts of England'.[123] Elsewhere, the Society explicitly turned down bids where there was felt to be insufficient railway accommodation. Hereford was one such casualty during deliberations in May 1852. At the time, the Worcester and Hereford Railway, the Newport, Abergavenny and Hereford Railway and the Hereford, Ross and Gloucester line were all being planned or constructed.[124] The promoters were acutely aware of the fact that 'Herefordshire could not continue to compete with other parts of the country unless it participated in the advantages of railway communication' and one of their number, Thomas Heywood, spoke for many of his fellow county landowners when he suggested that there was a feeling that, 'At present, they were completely in the background'.[125] Despite the scheduled opening of sections of these lines in 1853, RASE council members were unwilling to take on trust that railway access to Hereford would be completed and 'fit for use' by the time of its annual show in that year, and chose instead to visit Gloucester.[126] Deviations from the railway network, such as that to Shrewsbury in 1845, were a rarity, which had to be justified by the benefits that they might bring to an otherwise 'unenlightened' part of the country. This was certainly how RASE members liked to view the Shrewsbury show which, it was claimed, had brought more advantage to farming in the district in one week than it would otherwise have gained in 20 years.[127] In any case, its lack of railway connections were to some extent compensated by its good canal, road and river communications with Wales and the Midlands, and the fact that it could thus be regarded as 'the centre of a district'.[128]

Accessibility by rail was highly prized by the other associations too. Roderick Murchison backed the case for a BAAS meeting at Bath in 1844, 'for it is the centre of a network of rail roads, open to Ireland and the south-west and four

---

122 Goddard, *Harvests of Change*, pp. 32–3.

123 *Ipswich Journal*, 25 July 1863.

124 See for example, *Daily News*, 2 August 1851, 10 August 1852, 17 August 1852.

125 Landowner Thomas Heywood speaking at a meeting in the Feathers Hotel, Ledbury, quoted in *Daily News*, 18 July 1851.

126 MERL, SR RASE/B/I/4 Minutes of Council (rough copy), 5 May 1852.

127 This claim was made retrospectively by Chaloner. MERL, SR RASE/B/I/3, Minutes of Council (rough copy) 5 May 1847.

128 MERL, SR RASE B/1/2, Minutes of Council (rough copy), 1 May 1844.

hours from London'.[129] For towns which lay beyond the rail network to succeed in attracting the British Association, there had to be significant alternative attractions to compensate for the perceived inaccessibility. Such was the case when Cork was selected for the meeting in 1843. The success of the town in persuading the BAAS to visit in spite of the lengthy journey that would be faced by most delegates owed much to the strength of its cultural life and its range of local institutions. With well-established library and debating societies it had a reputation as 'the principal provincial hub for clubs and societies' in southern Ireland.[130] Five years later when the BAAS visited Swansea there was no railway connection, and delegates planning to travel from England were faced with an unpredictable sea passage from Bristol. But Swansea set great store on the claim to be a hub of scientific culture for Wales. With its recently established Royal Institution of South Wales, housed in a purpose-built museum building complete with lecture hall, it is possible to appreciate why the British Association overlooked the absence of a railway line for the chance to take its message of scientific enlightenment to Wales.

While such departures were justifiable, and sometimes necessary in the 1840s when the country was still far from fully served by the evolving rail network, by the 1870s things were different. New centres of demographic growth and commercial wealth were emerging and staking a claim, and older cities which had been among the first to play host to parliaments of science were returning to remind the associations of their attractions. Cardiff, for instance, which experienced rapid demographic growth in the second half of the nineteenth century, propelling it from fourth to first in the rank order of most populous Welsh towns between 1841 and 1871,[131] landed two annual meetings back-to-back in 1871 and 1872 with the RAI and RASE. For RASE, the choice of Cardiff marked a real change of heart since 1862 when an earlier application from the town was not even put to a vote.[132] A decade later, members of the RASE council noted that Cardiff was now 'very successful' and the support for the meeting was overwhelming.[133] Here there was a sense that the hosting of national scientific meetings marked the town's 'arrival' as a major urban centre. Bristol and Glasgow both came forward to present cases to the British Association council for the right to host its annual meeting in 1875. Dr Beddoe, for Bristol, pointed out that 'the south-western part of the country had not been visited by the Association for forty years and it was upon that ground extremely desirable that the Association should in 1875 pay a visit to that district'. (His claim was somewhat exaggerated: Exeter, Bath and

[129] Roderick Murchison to William Whewell, 1 March 1844, quoted in Morrell and Thackray, *Gentlemen of Science. Early Correspondence*, pp. 360–361.

[130] Clark, *British Clubs and Societies*, pp. 121–43.

[131] John Williams, *Digest of Welsh Historical Statistics*, vol. I (Cardiff, 1985), pp. 63–4.

[132] MERL, SR RASE/BI/6 Minutes of Council (rough copy), 7 May 1862.

[133] MERL, SR RASE/BI/7 Minutes of Council (rough copy), 3 May 1871.

Cheltenham had all hosted meetings since the last Bristol BAAS congress in 1836.) Glasgow's last BAAS meeting had been held 20 years earlier but, according to the Lord Provost, the 'immense strides' made by the town and its suburbs in the intervening years, and the 'chemical, engineering, and ship-building works, their mining establishments, and their own corporate undertakings were such as would more than interest those who paid a visit to their city'.[134]

With big players such as these coming forward to offer themselves as hosts it was little wonder that smaller towns sometimes became disillusioned about their prospects of being able to compete successfully for a parliament of science. It was a problem which led to the re-think in the hierarchy of the Royal Agricultural Society about the wisdom of its competitive method of selecting meeting venues. Members feared that the bidding process acted as a deterrent, and noted that, 'There are important towns at which it would be greatly to the advantage of the Society to hold meetings, which will not enter into a rival competition, whilst there are others which from having before incurred the expense and trouble of competing, and having been unsuccessful, will not again come forward'. Clearly there was a balance to be struck here between making the hosting of an association a desirable prize, but one that appeared to be within the reach of as wide a range of towns as possible. The decision of RASE to revise its method of selecting places for annual meetings resulted in its members taking on more responsibility for ascertaining information 'either privately or in a semi-official manner' about the towns they were interested in visiting so as to reduce the element of competition. It must have had the experience of places like Lancaster and Hereford in mind when it expressed the hope that, 'what has hitherto been a cause of great jealousy and disappointment will be removed and the interest of the Society thus promoted'.[135]

Table 2.2 Top 20 host towns, 1831–84*

Host Town	Number of Meetings
Newcastle-upon-Tyne	7
Birmingham	6
Edinburgh	6
Liverpool	6
Oxford	6
York	6
Bristol	5
Cambridge	5
Dublin	5
Glasgow	5

[134] *Leeds Mercury*, 23 September 1873.

[135] MERL, SR RASE, B/XIII/4, Miscellaneous, Special Country Meeting Districts Committee, 1 October 1876.

Manchester	5
Norwich	4
Plymouth	4
Southampton	4
Belfast	3
Canterbury	3
Carlisle	3
Exeter	3
Leeds	3
Shrewsbury	3

* Excludes London

The towns and cities which proved most successful in attracting parliaments of science in the mid-Victorian years were places which had a broad appeal and could tick a number of apparently contradictory boxes: populous and commercially or industrially successful, but not so new and unfinished that the infrastructure and buildings to support a meeting were lacking; well-established as a visitor venue and cultural centre, but not so distinguished in this respect as to render a meeting unnecessary; well-endowed with active town elites, but also well connected to neighbourhood gentry circles and the wider rural hinterland. Finding places which satisfied all of these counts was a difficult task, but a number of towns and cities did succeed in convincing several learned associations of their ability to do all, or most, of these things and, consequently, can be identified as congressional centres. The 20 towns which most frequently hosted the four associations featured in this study are listed in Table 2.2. Newcastle-upon-Tyne which, by 1884, had hosted seven meetings in all for the four associations was the most popular venue; Oxford, York, Edinburgh, Liverpool and Birmingham, each of which hosted six parliaments of science in the same period, followed close behind.[136] Alongside the university towns and the well-established trading ports came old regional centres like Norwich, which hosted each of the four associations once in the study period. Norwich's long-standing status as a centre for trade, textile manufacture and leisure, as well as its functions as a financial centre and assize town, gave it an enduring importance in its region. Indeed one commentator has noted that it stood out as a centre for culture and had no rival in East Anglia to its regional capital status before the twentieth century.[137] The ability of places like Norwich and York to compete successfully with, and in some cases to out-bid more populous industrial centres like Manchester and Leeds in the battle to host the major parliaments of science, was a measure of the resilience of the older county towns. The industrial metropolises of Victorian Britain found it difficult to compete with

[136] For a full list of host towns see Appendix 1, p. 165–66.

[137] Penelope Corfield, 'From Second City to Regional Capital', in Carole Rawcliffe and Richard G. Wilson (eds), *Norwich Since 1550* (London, 2004), pp. 139–66.

places where long-established institutions for leisure and learning offered ready-made facilities which could be adapted for use by visiting associations. Indeed for these older centres, the hosting of an association like the BAAS became a valuable means of re-asserting their importance at a time of significant urban and demographic change.

As barometers of town status, annual scientific meeting locations provide a kind of running commentary on the changing urban map of Britain in the mid-Victorian years. The deliberations of council members in the associations offer a contemporary eye view of urban rank based on their annual comparative assessments of the facilities and environment of the bidding towns. Architectural improvements, railway connections and commercial and demographic growth could transform a town's chances of selection, but unsuccessful bids could expose deficiencies in accommodation and communications, even in some large urban centres. The process of bidding for a meeting both fed on, and contributed to, the competitive urban climate of mid-nineteenth century provincial Britain. From the towns' perspective, entering into the bidding process was sometimes part of a wider quest for status and recognition ahead of other towns. In other words, there was much more at stake than simply the hosting of a week-long scientific event. These meetings presented towns with an opportunity to attract visitors, generate income, command press attention and showcase some of their best buildings and institutions. But success in the bidding process was just the beginning. The real work of generating cultural capital for the host town began once the work of organising and running the congress got underway.

Chapter 3
Running the Meeting

'A fortnight hence, on the heels of the equinoxial gales, and abreast with pheasant shooting and Michaelmas geese, the National Association for the Promotion of Social Science will hold its tenth annual meeting in this city'. So wrote the editors of the *Manchester Examiner* in September 1866 in anticipation of the appearance of 'the sciences and humanities … before us in bodily shape'.[1] Annual meetings typically lasted a week and had the potential to draw in thousands of visitors: the Social Science Association's annual gatherings in Edinburgh and Glasgow in the late 1850s, for example, attracted around 2,800 delegates.[2] The normal rhythms and patterns of urban life were suspended for the duration. Public rooms, school buildings, theatres, chapels and law courts were diverted from their normal functions to provide room for lecturers and their audiences. So that all energies could be focused on the events of the week, like-minded societies put their own activities on hold, as in Worcester when the local agricultural society abandoned its own show in 1863 so that effort and attention would not be deflected from the arrival of the Royal Agricultural Society of England. It was, according to one reporter, 'an instance of sinking the local in the imperial'[3] and a tacit acknowledgement of the greater rewards to be gained by engaging with a national organisation.

Tightly timetabled events programmes unfolded over the ensuing days, maximising every opportunity for information to be presented and exchanged and for enthusiasts and experts to meet, mingle and converse. The Social Science Association's annual meeting at Liverpool in 1858 is a representative example, commencing on Monday 11 October with a church service at St Nicholas', prayers by the Rector of Liverpool and a sermon by the Lord Bishop of Chester, followed by a general meeting in the evening, addressed by Lord John Russell, in St George's Hall. On Tuesday, addresses by the Lord Chancellor of Ireland and the Earl of Carlisle were held in the concert room of St George's Hall and later delegates could take part in an excursion to the Akbar Reformatory Frigate[4] before attending further addresses in the concert room by the Earl of Shaftesbury and Sir James Stevens. Evening entertainment consisted of an art exhibition at the Liverpool Academy, a concert by the Philharmonic Society and a soirée hosted

---

1 MCL, BR F306 S3, *Manchester Examiner*, 20 September 1866.

2 Goldman, *Science, Reform and Politics*, p. 93.

3 *Birmingham Daily Post*, 15 July 1863.

4 A training ship, moored on the Mersey from 1855, for the discipline and instruction of young Protestant offenders. See Michael Macilwee, *The Liverpool Underworld: crime in the city, 1750–1900* (Liverpool, 2011), p. 49.

by the mayor. On Wednesday, Thursday and Friday, the various departments, or sections, of the Association met in separate meeting rooms in St George's Hall between 10.30 and 4pm for papers and discussion. Evenings were also utilised for an address by Lord Brougham on the Wednesday, a soirée for members and associates on the Thursday and a public dinner in St George's Hall on the Friday. On Saturday, the business of the week was concluded with a final meeting in the concert room with reports from each of the sections and closing resolutions.[5] This formula of business meetings, learned exchanges, outings and entertainments was followed in the annual meetings of the BAAS, the RAI and the Social Science Association. RASE events, centred on an outdoor showground, also followed a prescribed pattern of judges' inspections and implement trials, of the opening of the machinery yards and cattle enclosures to the public, of dinners for RASE council members and of a mass pavilion dinner.[6]

These week-long scientific meetings were months in the planning. In each case, the receiving town was, to a large degree, entrusted to make all of the necessary local arrangements, from finding and equipping suitable meeting rooms or a showground, to securing residential accommodation and organising the hospitality and catering for the event. In Newcastle, in 1863, the local organising committee comprising an executive and sub-sections to organise the finance, entertainments and lodgings, met a total of 79 times between February and November.[7] Preparations were carried out under the watchful gaze, and sometimes the direct intervention, of the parent association, whose reputation rested on a successful outcome. Ongoing dialogue between central figures from the national association and local activists during months of planning was of a different tenor from the communications during the bidding phase: no longer was the town in the position of petitioner; now it was a partner in the business of knowledge dissemination with ambitions to shape both the science and the sociability of the annual meetings. But the boundary between central control and local autonomy in the running of the annual meetings was a contested one. The leeway given to local organising committees in making the arrangements was not unlimited. Sometimes this gave rise to tensions, misunderstandings and argument if the controlling oversight of the national association was seen to be challenged too directly, or if finances on either side were strained. But, more often than not, the mutual reliance of town and visiting organisation on one another, and the vested interest of each in a successful meeting, was sufficient to ensure an effective partnership.

As well as providing what the visiting association required, organisers had their own ambitions for meeting week. There was enormous scope to use the occasion to show off the attributes of the host town to the outside world. In

5 NML MAL, John Towne Danson, D/D/III/17/5a, printed programme of the NAPSS second annual meeting, Liverpool, 1858.

6 *Derby Mercury*, 6 July 1881.

7 *Newcastle Courant*, 13 November 1863.

particular, the hospitality, entertainment, outings and dinners, in public and private venues throughout the presiding town, gave municipal leaders, local dignitaries and other members of the urban elite a platform on which to display what they perceived to be their own strengths and assets. To do this they were quite prepared to move well beyond what was required of them by the visiting organisation. Much organisational energy and expense was channelled into the publication of special volumes or histories of the town, the issuing of maps or guides highlighting places of special interest, the opening up of local institutions and manufacturing sites for inspection by guests and the choice of excursions, exhibitions and displays. As well as offering some light relief to delegates from their scientific pursuits, these elements were designed to draw attention to the best features of the natural, cultural and urban environment of the town and represented ways in which the town put itself on display. The scientific content of the meetings, too, could be tailored to reflect the location through the inclusion of papers which examined local flora, fauna, geological, social or archaeological features or, in the case of RASE, the showing of local livestock breeds. Initiatives to extend national standards and stipulations over regional practices of scientific observation and information gathering were a feature of the mid-Victorian period. As other studies have shown, the harnessing of local expertise for the advancement of a national scientific project involved negotiation and acceptance that personnel and procedures in different localities did not always operate according to centrally prescribed regulations.[8] In the case of the parliaments of science, the local slanting of knowledge content was often done at the instigation of the association, whose leaders concerned themselves with the attraction of something novel and unique in each meeting location as a means of securing high attendance and netting a good income from ticket sales.

Above all, host towns used the most powerful tool at their disposal to put their own indelible mark on parliaments of science when they applied customary forms of civic ceremony and traditional corporation ritual to the occasion. Civic ceremonies were just one of a host of urban rituals which were played out in provincial towns in the post-Restoration period.[9] Research on civic ceremonies in urban Britain since the seventeenth century has traditionally focused on public celebrations such as coronations, monarch's birthdays and royal visits. Such occasions afforded welcome opportunities for town leaders to don their robes of office and make a public display of loyalty and leadership through dining, toasting, processing and speech-making while controlling and containing the activities of the lower orders. By the 1820s, these occasions had become strained by socio-

[8] See for example, Simon Naylor, 'Nationalizing provincial weather: meteorology in nineteenth-century Cornwall', *British Journal for the History of Science*, 39/3 (2006): 407–33.

[9] Peter N. Borsay, '"All the town's a stage": urban ritual and ceremony, 1660–1800', in P. Clarke (ed.), *The Transformation of English Provincial Towns, 1600–1800* (London, 1984), pp. 228–58.

economic change. Popular participation declined as urban leaders sought to reduce opportunities for unruly revelling leading, sometimes, to resentment and protest.[10] These traditional celebrations became associated with an old, unreformed order and with political and social division. But research on civic ritual has stressed its capacity for adaptation and re-invention,[11] and as this chapter will show, the great meetings of the BAAS from the 1830s, and the other national knowledge associations formed in the next two decades, offered a new arena in which civic ceremony could be played out. Research on provincial scientific institutions in nineteenth-century Britain has highlighted a civic dimension to their activities,[12] but the way in which civic ritual was used at major scientific meetings has not, hitherto, been explored. Here it will be shown that town dignitaries used the full range of measures at their disposal, from gift-giving, luncheons, robe-wearing and displaying of corporation symbols like maces and swords, the reading of welcoming addresses and the decoration of the streets. As a result, the great anniversary meetings of the national knowledge associations of the mid-nineteenth century became closely interwoven with patterns of civic ritual in Victorian towns. Their aim, it will be suggested, was not only to communicate a generous welcome and hospitality to visiting associations, but also to present a united front in support of the cause of science and learning. The provision of luncheons and other civic receptions were a particularly useful means of bridging factional divides as townsmen who usually moved in different circles came together in their capacity as hosts. By deploying all the pomp and paraphernalia of civic ceremony, host towns complemented the elements of spectacle and display in mid-Victorian science,[13] to create occasions which commanded widespread interest and attention.

Contemporary press reports necessarily loom large in providing the core evidence of the scientific and social business of the annual meetings. Provincial and national newspapers produced voluminous reports on the lectures, discussions, excursions and dinners of meeting weeks and described in great detail the appearance of the indoor and outdoor spaces in which knowledge exchange took place. Host towns and national associations were equally anxious to secure good press and did not leave matters to chance. Organisers went to considerable lengths to ensure that newspaper correspondents were fully briefed and given appropriate access to the key events of the congress. Maximum press coverage

10 George A. Tresidder, 'Coronation Day Celebrations in English towns, 1685–1821', *British Journal for Eighteenth Century Studies*, 15/1 (1992): 1–16.

11 David Cannadine, 'The Transformation of Civic Ritual in Modern Britain: the Colchester Oyster Feast', *Past and Present*, 94 (1982): 107–30.

12 See for example, Charles W.J. Withers and Diarmid A. Finnegan, 'Natural history societies, fieldwork and local knowledge in nineteenth-century Scotland: towards a historical geography of civic science', *Cultural Geographies*, 10 (2003): 334–53.

13 See for example, Iwan R. Morus, 'Worlds of Wonder: sensation and the Victorian scientific performance', *Isis*, 101/4 (2010): 806–16.

could be achieved by supplying the content matter for articles direct to as many newspaper editors as possible. The local organisers of the Manchester Social Science meeting in 1866 sent a programme summary to 40 local and regional newspapers on 26 September, a week in advance of the opening of the congress.[14] The tireless George Hastings also did his best to ensure that newspaper journalists were given every opportunity to report extensively on NAPSS meetings. In 1864, when the congress was due to be held at York, he anticipated delays in press coverage because 'There are only weekly papers at York and they would not be able to set up in the same rapid way as daily papers'. His solution was to request an advance copy from Lord Brougham of his presidential address to dispatch to key provincial organs including the *Leeds Mercury* and the *Newcastle Chronicle* so that their reports could be more swiftly compiled.[15] Elsewhere, newspaper proprietors issued special editions to supply themselves with extra column space to cover the proceeding of annual meetings. The editor of the *Bath Chronicle* announced to his readers that a special daily edition of the newspaper would be issued in addition to the regular weekly copy during the visit of the British Association to the town in 1864. The new daily edition contained full reports of lectures, addresses and discussions and was later reprinted as a single volume.[16] In Leicester, meanwhile, the RASE meeting of 1868 prompted the proprietor of the *Chronicle* and *Mercury* to publish a free supplement on Saturday 18 July consisting of 24 columns of coverage of the agricultural and horticultural shows along with associated advertisements.[17]

The reliability of these accounts cannot be taken for granted. Journalists were quick to deride the efforts of the learned associations where these were judged to be insufficiently 'scientific'. The Social Science Association in particular, felt itself ill-served by the London press,[18] and the demise of the British Association for the Advancement of Science was predicted on more than one occasion in *The Times* in the wake of annual meetings which were deemed to be either too modestly attended or swamped with people more interested in entertainment than educational insight.[19] Reports in provincial newspapers, too, could be loaded with regional rivalries or political partisanship, as evidenced in the Tory *Blackburn Standard*'s scathing account of the excursion to the Cobden Memorial Cotton

---

14 MCL, BR F306 N1, National Association for the Promotion of Social Science, circulars etc.

15 UCL Special Collections, B MSS 13,028, 6 September 1864; and 13,029, 8 September 1864.

16 Bodleian Library, Dep. BAAS 147, *The British Association for the Advancement of Science, Bath 1864: authorised reprint of the reports in the special daily editions of the Bath Chronicle* (Bath, 1864).

17 See notice issued in *Leicester Chronicle* and *Leicestershire Mercury*, 13 June 1868.

18 See for example, UCL Special Collections, B MSS 13,070, George W. Hastings to Lord Brougham, 8 December 1856 and 13,014, Hastings to Brougham, 12 August 1862.

19 See for example, *The Times*, 19 August 1835; 25 June 1845.

Mill, organised as part of the Social Science congress in Manchester in 1866.[20] But, as other research has revealed, local newspapers, even where these were of differing political persuasions, tended to unite in support of the efforts of local organisers when learned associations visited their own town.[21] There was a conscious setting aside of differences while under the scrutinizing gaze of outsiders. Ahead of the 1866 visit of the Social Science Association, the editors of the *Manchester Examiner* spelled out their intention to 'waive for three weeks at least, the part of critics, and straightway begin our task as hearty coadjutors'.[22]

This willingness to suspend hostilities and present a united front when a learned association was in town was a feature of meeting week in many locations. The day before the opening of the British Association for the Advancement of Science's annual meeting in Bristol in 1836, a council resolution was passed expressing:

> Their thankful appreciation of the honour and advantage which this city has enjoyed in being the seat of this anniversary meeting by which so many gentlemen of great scientific eminence have been brought within the personal observation of the residents … [and] of the delight they have experienced in the absence of everything inconsistent with the rights of the neutral ground of science.[23]

As their city moved from a period of political agitation and reform riot to its first municipal election of the reformed era,[24] it was small wonder that Bristolians derived such satisfaction from an occasion which had none of the political strains of earlier years but offered instead a cause around which all could unite. It was an early, but not an isolated example of the value towns placed on science as a 'neutral', unifying influence, and it demonstrated one of the benefits which towns expected to derive from hosting a scientific congress. Similar sentiments were expressed by other corporate bodies in resolutions expressing their thanks to the visiting association for choosing them to host and their dedication to the cause of knowledge and improvement. But the reading of these resolutions was often as much about articulating the town's view of itself as host as it was about voicing welcome or thanks to visitors. At Newcastle upon Tyne in 1852, the Royal Archaeological Institute was formally received by the right worshipful

20 See *The Blackburn Standard*, 10 October 1866.

21 Sherra Murphy, 'The learned gentlemen are in town: the British Association meeting of 1857 in Dublin's popular press', in Juliana Adelman and Éadaoin Agnew (eds), *Science and Technology in Nineteenth-Century Ireland* (Dublin, 2011), p. 164.

22 MCL, BR F306 S3, *Manchester Examiner*, 20 September 1866.

23 BRO, 32079 (39) Minute Book of the reception committee for the British Association visit, 26 August 1836.

24 Steve Poole, 'To be a Bristolian: civic identity and the social order, 1750–1850', in Madge Dresser and Philip Ollerenshaw (eds), *The Making of Modern Bristol* (Tiverton, 1996), pp. 85–6.

mayor, James Hodgson esq., the aldermen and the councillors of the borough. The town clerk, John Clayton, read the following address:

> We the mayor, aldermen and councillors of the borough of Newcastle upon Tyne, offer to you the president and members of the Archaeological Institute of Great Britain and Ireland, the sincere expression of our feelings of gratification on the occasion of your visit to the ancient town of Newcastle upon Tyne … Occupied as we are in the pursuits of commerce, we are at the same time sensible of the value of those of literature and science in which men of all parties and of all opinions can unite with one common object, to civilise and instruct mankind.[25]

This idea of the learned meeting as an opportunity to promote unity was more than just empty rhetoric. In a number of places, townsmen seem to have viewed a week of educational business as an opportunity to set aside differences of class or creed. Indeed co-operation was essential if the organisational feat of hosting a major congress was to be achieved. In Cheltenham, a town riven by sharp party political divisions and religious controversy,[26] the Social Science congress of 1878 promoted this greater sense of unity. The conversazione held at the Ladies College, 'facilitated that blending of the sections of local society' by bringing people together under one roof to entertain their visitors and to join in their appreciation of the exhibitions and entertainments on offer. In a place which was normally 'split up into little cliques', the 'neutral' presence of the visiting social scientists helped blur the usual lines of division.[27] Similarly, in Liverpool in 1858, members of the local organising committee who had made all the arrangements to host the National Association for the Promotion of Social Science that year put the success of their efforts down to a spirit of co-operation and inclusiveness: 'All parties, denominations, all views were represented, and each member of the committee had gone into his own circle, [and] inculcated an enthusiasm for the cause'.[28] In a port town noted for its sectarianism, and where 'all departments of local administration became a battleground in the religious conflict',[29] such co-operation was crucial, but a level of outside intervention was sometimes needed in order to make sure it was achieved. In 1837, John Phillips, BAAS secretary, had intervened in the composition of the Liverpool organising committee responsible for making arrangements for the visit of the BAAS to the city that year: 'Jealousies as to political and other disunion have induced me to recommend … that the local officers should NOT constitute the Local Council,

---

25 SAL, WAY MS 790, RAI Meeting at Newcastle, 1852.

26 Gwen Hart, *A History of Cheltenham* (Leicester, 1965), pp. 211–57.

27 *Bristol Mercury and Daily Post*, 31 October 1878.

28 NML MAL, John Towne Danson, D/D/III/17/5a: Newspaper Cutting Book, *Northern Times*, 15 November 1858.

29 Frank Neal, *Sectarian Violence. The Liverpool Experience, 1819–1914. An Aspect of Anglo-Irish History* (Manchester, 1988), p. 177.

but with care and judgment select names of leaders of institutions and public or influential bodies, who are likely to work, and to get these declared as Local Council by the London Council'.[30] Elsewhere, too, rivalries were put on hold, even for the good of a successful congress in a neighbouring urban centre. In 1860 George Hastings reported approvingly to Lord Brougham that preparations for the forthcoming Social Science congress in Glasgow were being aided by Edinburgh: 'I am glad to say that the Edinburgh people have taken up the matter heartily, discarded (if they ever entertained) any foolish jealousy at Glasgow, and have established a branch committee to aid the meeting'.[31]

It was in the communication between association representatives like Hastings and the members of the organising committees in the host towns that the nuances of responsibility and control of the annual meetings were played out. Aware of how important the talents of the local men were to the success of the meetings, but also anxious to keep a tight rein on arrangements, association secretaries employed a mixture of consensus and compulsion in their dealings with organising committees. Hastings, for example, came up with a face-saving compromise when the Manchester committee proposed their own candidate for the presidency of the Social Science congress in the town in 1866. After reminding them that 'the appointment of President lay with the Council', he offered to put their suggestions before the association, although he later wrote to Lord Brougham and suggested that he make the same proposal for president: 'The nomination will then come from you and from no one else and the Manchester resolution, which is certainly beyond their functions, may be passed over'.[32] Such exchanges were illustrative of the relationship between the provincial urban centre and the national organisation as one of mutual dependence but not of equals. Sometimes the abilities of local officials gave cause for concern. At York in 1864, the commitment of the Lord Mayor to the success of the Social Science meeting was described by Hastings as 'luke warm'. He later reported to Brougham that 'his conduct is attributed more to silliness than anything else which I should imagine from his appearance and manners'. He had no doubt that the mayor 'may be managed or put down'.[33] But elsewhere, the efforts of very able town officers were praised. Roderick Murchison, who visited Birmingham to meet the local organising committee for the 1839 BAAS meeting, reported enthusiastically to Harcourt on the quality of its membership:

---

30 John Phillips to William Harcourt, 6 April 1837, quoted in Morrell and Thackray (eds), *Gentlemen of Science. Early Correspondence*, p. 240.

31 UCL Special Collections, B MSS, 13,002, George W. Hastings to Lord Brougham, 17 April 1860.

32 UCL Special Collections, B MSS, 13,150, George W. Hastings to Lord Brougham, 14 June 1866.

33 UCL Special Collections, B MSS 24,820 and 13,026, George W. Hastings to Lord Brougham, 22 March and 1 September 1864.

> Mr Barker is an eminent solicitor and the most gentlemanlike man in the place ... Mr Hodgson is [the] leading surgeon, FRS, and a very pleasing, well-informed, right-minded man, who takes no care of politics. Mr Follett Osler is the clever, active youth who invented the anemometer; he is a violent radical in politics. Dr Peyton Blakiston is an incipient physician of the conservative order ... Mr Moilliet, the Local Treasurer, is a pleasant old Swiss man who has made a large fortune here; and Mr Russell, the permanent Local Treasurer is a zealous, indefatigable man, ready to slave at anything.[34]

With serious Chartist unrest in the town just a month earlier,[35] the BAAS had good reason to vet the credentials of the Birmingham organisers but, as elsewhere, the participation of men of different political persuasions in the organising committee illustrated the capacity of the scientific congress to transcend party rivalries.

Town dignitaries played a prominent role in the reception of learned visitors and the inauguration of their annual meetings. Towns large and small mustered a display of civic hospitality which provided both a formal welcome for distinguished guests and an opportunity for local officials, with all the trappings of their office, to assemble alongside the leading lights of the country's foremost knowledge associations. At Hull, where the British Association meeting was held in 1853, objects reflecting the history of the corporation were laid out for display at the Music Hall during an evening reception. 'The mayor promenaded in graceful dignity, and bearing his ponderous chain of office, pointed out to visitors all the paraphernalia of civic pomp, including swords and tankards presented to the corporation by Henry VIII'.[36] For a relatively small town, the opportunity to draw attention to its civic credentials while basking in the national gaze was too good to miss.

At Bury St Edmunds, the arrival of the Royal Archaeological Institute in July 1869 was marked by an official, but relatively private and small-scale reception. The members of the Institute were 'received in full corporate dignity at the Town Hall' where the mayor, 'dressed in full mayoral robes, and preceded by the mace bearers and members of the corporation', took to the platform, followed by a string of speakers representing the local archaeological society, the local clergy and the Suffolk gentry, who addressed the gathered audience of about 200. A 'sumptuous lunch' was then served at the mayor's invitation at the nearby Angel Hotel.[37] Royal Archaeological Institute meetings were especially appropriate occasions on which to display the antiquities of local government. At Cardiff in 1871, exhibits on show in the temporary museum, housed for the week in the Assembly Room at

---

[34] Roderick Murchison to William Harcourt, 8 November 1838, quoted in Morrell and Thackray, *Gentlemen of Science. Early Correspondence*, pp. 292–3.

[35] For details see, Conrad Gill, *History of Birmingham, Volume I: Manor and Borough to 1865* (London, 1952), pp. 240–271.

[36] *Morning Chronicle*, 10 September 1853.

[37] *Ipswich Journal*, 24 July 1869.

Figure 3.1 A civic welcome for British Association members at Swansea in 1880, *The Graphic*, 4 September 1880 (author's collection)

the Town Hall, included a silver mace and the original charter of the Borough of Llantrisant.[38] At Plymouth in 1877, an annual corporation ceremony was utilised as an excursion for delegates attending the British Association meeting in the town that year. The wife of geologist William Pengelly described the proceedings in a letter to her sister:

> Today William and several other members of the Association are gone to attend a curious old ceremony, held by the Corporation every year in commemoration of Sir Francis Drake having brought in the water here from the moor ... The ceremony is held about nine miles from here. It consists in first drinking to the memory of Sir Francis Drake in water, and then comes the Toast, 'May the descendants of him who gave us water never want wine'. The first fish dish served always consists of fish caught in the Leat.[39]

Such private shows of corporation ritual and hospitality allowed a close-knit elite group to fraternise with the visitors and fostered a sense of privilege. Where the urban setting was more diverse, civic receptions had to be opened up so as not to exclude or offend would-be participants. The opening day of the NAPSS' sixteenth annual congress at Plymouth in 1872 included a 'procession of the Corporations of Plymouth and Devonport to St Andrew's Church'.[40] A church service was the

38 *The Archaeological Journal*, XXVIII (1871), pp. 337–8.

39 Mrs Pengelly to her sister, Plymouth 24 August 1877, in Hester Pengelly (ed.), *A Memoir of William Pengelly of Torquay, FRS, Geologist, with a Selection from His Correspondence* (London, 1897), p. 251.

40 *Devon Weekly Times*, 13 September 1872.

traditional opening to a Social Science congress,[41] but the processional element provided scope for a public display of civic endorsement which was particularly important in this instance because Plymouth was hosting the Association on behalf of the wider urban conurbation comprising the 'three towns' of Plymouth, Devonport and Stonehouse.[42] Similar issues arose in Birmingham when the British Association visited in 1865, as it had done on two previous occasions. The desire of the local organising committee to take their guests on excursions in south Staffordshire brought municipal leaders in neighbouring towns into the frame. In Dudley, for example, the mayor convened a meeting to plan a civic welcome for BAAS members visiting the Earl of Dudley's ironworks at Round Oak. Rather than preside over hospitality there, the mayor insisted that 'his duties should, during his year of office, be strictly of a civic character, and that his fellow townsmen had clearly a claim upon his attention'. His suggestion, instead, was that the visitors, who it was hoped would include Sir Roderick Murchison, should be publicly invited to a luncheon where, 'if the proceedings assumed a public character their own public men must be present and take part'.[43] The gesture was clearly designed to garner the attention of neighbouring towns as much as it was to cater for the needs of the visiting scientists, with the mayor noting approvingly that his plans had aroused a sense of jealousy in Wolverhampton.[44]

If the prospect of eminent scientists arriving was enough to stir civic leaders into action, the occasions when royal guests were also among the meeting delegates raised the ceremonial stakes still further. Such was the situation in Southampton when Prince Albert attended the British Association meeting there in 1846. The Prince arrived in the town on the Monday of the meeting week and attended a number of the sections, thus placing the town corporation and the hierarchy of the British Association into the position of joint hosts of the royal party. The day was carefully choreographed and stage-managed with the town mayor and the president of the Association, Sir Roderick Murchison, sharing the hospitality duties. From an early hour:

> Flags were waving in all direction, and a number of the neighbouring gentry came into the town. At half past ten o'clock the mayor, aldermen and common council, assembled in their robes in the reading-room, at the utmost extremity of the pier, close to the principal landing-place. They were soon afterwards joined by Sir R. Murchison, to whom the mayor introduced the several members of the corporation. The room had been handsomely fitted up and decorated, and a carpet was laid down to the water's edge.[45]

---

41 Goldman, *Science, Reform and Politics*, p. 95.

42 Crispin Gill, *Plymouth. A New History* (Tiverton, 1993), pp. 225–42.

43 *Birmingham Daily Post*, 22 August 1865.

44 Ibid.

45 *Hampshire Telegraph and Sussex Chronicle*, 19 September 1846.

Following his arrival at the pier, the Prince was conveyed in the mayoral carriage, accompanied by both the mayor and Murchison, to the Bugle Hall where Section A was in session. There, the president of the section, Herschel, presented him with the gift of a coloured geological drawing of the Isle of Wight, prepared by a Southampton man, Mr Drew, one of the secretaries of Section A.[46] It was the perfectly conceived memento, representing both the science and the region, as well as the talent of a local geological artist.

After his initial appearance at the Southampton meeting, Albert became a regular attender at BAAS events. The following year at Oxford, he visited all of the sections and lunched with some 150 British Association members.[47] The presence of royalty helped cement connections in contemporary minds between the hosting of congresses and more traditional ceremonial occasions in the towns. The news that Prince Albert would attend the British Association meeting at Ipswich in 1851 led to some fond recalling of earlier royal visits to the town:

> … in ancient times, it is true, the town of Ipswich was honoured by Royal visits, the public records bear upon them the names of Charles II and Queen Elizabeth; and at a comparatively modern date, HRH the Prince Regent, afterwards George III, made his public progress through our loyal town. But none of those occasions could exceed in interest the present most auspicious event.[48]

Elsewhere, the festivities which greeted a visiting association were compared with recent royal celebrations. At Worcester in 1863, the arrival of the Royal Agricultural Society of England for their annual meeting in July was greeted with 'a display of flags and illuminations which remind us of the 10th of March rejoicings'.[49] On 10 March 1863, the marriage of the Prince of Wales had been celebrated enthusiastically in Birmingham and other Midlands towns, particularly with the use of flags 'flying over the fronts of the buildings – displaying their bright colours, indeed in all directions and in every conceivable form'.[50] At Swansea when the BAAS visited for the first time in 1848, several of the ships in the bay 'had colours flying, as on a royal visit'.[51] That contemporaries drew such comparisons illustrates the connecting threads between forms of royal ceremonial and the visits of the great nineteenth-century knowledge associations to provincial towns and cities.

---

46 Ibid.

47 Letter from Adam Sedgwick to Isabella Sedgwick, 12 July 1847, in John W. Clark, *The Life and Letters of the Rev. Adam Sedgwick*, vol. II (Cambridge, 1890), p. 123.

48 *Ipswich Journal*, 5 July 1851.

49 *Birmingham Daily Post*, 20 July 1863.

50 *Birmingham Daily Post*, 11 March 1863.

51 *Morning Post*, 11 August 1848.

The new levels of pageantry stimulated in Britain's large industrial cities by Queen Victoria's visits in the 1850s[52] seem to have infected the work of local organising committees preparing for the arrival of visiting associations. By the 1850s, external decorations, particularly the use of flags, bunting and evergreens, to festoon the streets of the host town during meeting week was becoming the norm, to the extent that towns which fell below expectations in this regard faced criticism in the press.[53] This was especially so in the case of the Royal Agricultural Society of England, whose shows were increasingly accompanied by displays of festive decoration, not just in the showground itself, but in the major streets and thoroughfares of the town. At Chester in 1858, the organisers established a decorations committee to oversee and direct the ornamentation of the town for the duration of the RASE visit.[54] Elsewhere, private residents and commercial enterprises were encouraged to take the initiative. This RASE tradition permeated through to the county agricultural societies, some of which offered prizes to encourage local residents and businesses to adorn their premises with festive displays during show week.[55] In Worcester, where the Royal Agricultural Society show was held in 1863, some of the most striking embellishments were those at the Railway Station in Foregate Street where, 'the company's officers have erected two triumphal arches and decorated the bridge which crosses the street. A very tasty evergreen arch spans the cross bearing the inscription, "welcome to the R.A.S.E." on each side; near the Cathedral is another, bearing like expressions of welcome'.[56] It was apposite that the railway station was a special focus for decoration. Lack of adequate railway communication had snuffed out Worcester's chances of hosting the Society ten years earlier.[57] Now, with the help of these elaborate displays, the point could be made that railway transport lay at the heart of the urban infrastructure. As this case suggests, the value of street adornments was more than purely aesthetic. If tactically deployed, garlands, arches, flags and other decorative devices could be used to create focal points in the streetscape, and to draw attention to the best assets of the town.

Triumphal arches marking processional routes or decorating key thoroughfares were a familiar symbol of welcome for royal visits to Victorian towns.[58] Their

---

52 Simon Gunn, 'Ritual and Civic Culture in the English Industrial City, c.1835–1914', in Robert J. Morris and Rick H. Trainor (eds), *Urban Governance. Britain and Beyond Since 1750* (Aldershot, 2000), p. 228.

53 Such was the case at Lincoln, see *Morning Post*, 19 July 1854.

54 *Chester Observer and General Advertiser*, 17 July 1858.

55 See for example, Andrew Walker, 'The Lincolnshire Show, 1869-1913: rural and urban encounters on the showground', in Keith Grieves et.al. (eds), *Rural and Urban Encounters in the Nineteenth and Twentieth Centuries: Regional Perspectives* (Conference of Regional and Local Historians, 2004), pp. 46–7.

56 *Birmingham Daily Post*, 20 July 1863.

57 *Ipswich Journal*, 25 July 1863.

58 See for example, Margarita Cappock, 'Pageantry or Propaganda? The *Illustrated London News* and Royal Visitors in Ireland', *Irish Arts Review Yearbook*, 16 (2000): 86–93.

Figure 3.2 Triumphal Arch in Tavern Street, Ipswich, for the 1851 meeting of the British Association for the Advancement of Science, *Illustrated London News*, 12 July 1851 (author's collection)

use during annual meetings was most common where royal guests were among the delegates, as at the Ipswich BAAS meeting in 1851 with Prince Albert in attendance: 'At the corner of every street, wherever his Royal Highness was to pass, or even where it was supposed he might command a view of it from his carriage, a triumphal arch was erected, lavishly adorned with evergreens and flowers'.[59] Where the opportunity presented itself to use a second-hand arch, previously employed for a royal occasion, towns were not averse to adapting these for use during annual meetings. At Leeds in 1858, the British Association meeting followed in close succession to the Queen's visit to inaugurate the new town hall and the triumphal arch erected at the top of East Parade to mark the arrival of her majesty was left in place for the duration of the meeting.[60] This use of royal visit paraphernalia at annual congresses, as well as the occasional

59 *Morning Chronicle*, 4 July 1851.

60 *Leeds Mercury*, 2 October 1858.

attendance of royal guests in person, all helped bring, in the words of one eminent scientist of the day, 'an éclat to the meeting'.[61]

While street decorations were used to embellish key urban landmarks in the host town, great attention was also paid by local organisers to the arrangement and appearance of the interior spaces used during meeting week. The need for a general reception area for meeting delegates was the first challenge faced by the host venue. In Cheltenham in 1856, the Montpellier Rotunda was used as the place where, 'all parties must, in the first instance, of necessity, proceed to pay their subscriptions, procure tickets of admission to the meetings, and the reports and journals … and to which place also all the letters and packages of the members are addressed'.[62] At Bradford, where the British Association meeting was held in 1873, the Exchange Hall performed this function and was praised as a 'first-class reception room' with telegraphic wire and temporary post office available on-site. To assist visitors with their enquiries, a series of booths was set up around the hall, each designated to deal with a different item of business. Organisers were commended for the 'completeness and efficiency' of these arrangements by which information was provided to the throngs of people arriving with enquiries on the first day.[63]

Basic decorating such as the white lining, painting and fitting up of these areas and the furnishing of meeting rooms with benches and baize cloth were all undertaken by local organisers.[64] But as well as functionality, the aesthetic appearance of the spaces in which large numbers of visitors were due to assemble was of the utmost importance. Through careful use of lighting, plants, pictures and other decorations, organisers could convey an image of their town as a place of modernity and refined taste. At the NAPSS congress in Manchester in 1866, local press reports drew attention to the appearance of the Central Hall decked out for a soirée. It was 'well lighted' and 'was set throughout in admirable taste with lounges and settees, the build and fashion of which harmonised with the character of the hall. The simplicity of the decorations was another evidence of good taste'.[65]

But local committee members were also alert to the importance of spectacle to scientific audiences in the Victorian period. Where possible, they created visual effects which were designed not just to decorate but to inspire a sense of wonder among onlookers. At Bury St Edmunds in 1867, when the Royal Agricultural Society was in town, visitors were treated to a show of illuminations in the evenings, including the use of a magnesium light, projected from the top floor of the Angel Hotel onto the Abbey Gate 'with beautiful effect, the minutest detail

---

61 Letter from Adam Sedgwick to Isabella Sedgwick, 12 July 1847, in Clark, *The Life and Letters of the Rev. Adam Sedgwick*, vol. II, p. 123.

62 *Morning Chronicle*, 30 July 1856.

63 *Daily News*, 17 September 1873.

64 See for example, BRO 32079 (39) Minute Book of the Reception Committee for the British Association visit, 27 July 1836; 5 August 1836.

65 MCL, BR F306 S3, *Manchester Guardian*, 4 October 1866.

Figure 3.3 Reception area in the Exchange Assembly Room for British Association delegates at Birmingham in 1865, *Illustrated London News*, 23 September 1865 (author's collection)

of its exquisite architecture being shown with the utmost distinctness'.[66] The use of gas lighting, in particular, though no longer novel,[67] created some attractive decorative effects which drew admiring comment from visitors. At Newcastle in 1838, the Green Market was lavishly decorated for a promenade:

> The floor was entirely covered with canvass, and the pillars were admired with festoons of evergreens, intermixed with flowers. The fountains played during the whole of the evening, and a fairy effect was produced on the falling water by the circle of the fountain being lit up with gas, over which it fell.[68]

The scene made a favourable impression on John Phillips, visiting prior to the meeting in his capacity as BAAS assistant secretary, who described it as 'truly one of the most marvellous rooms ever beheld'.[69]

---

66 *The Bury and Norwich Post* and *Suffolk Herald*, 23 July 1867.

67 Chris Otter, *The Victorian Eye. A Political History of Light and Vision in Britain, 1800–1910* (Chicago, 2008), pp. 135–72.

68 *Newcastle Courant*, 24 August 1838.

69 John Phillips to William Harcourt, 10 August 1838, in Morrell and Thackray (eds.), *Gentlemen of Science. Early Correspondence*, p. 273.

The creation of such attractive displays did not come cheap, and the cost of decoration was one of a number of key expenses which placed a heavy financial burden on the host town. At Newcastle, the lighting of the Green Market was itemised along with the fitting up the Corn Market and the Circus among the principal costs that the town would have to bear to stage the BAAS meeting.[70] When news reached the town that the BAAS visit had been secured, voices of caution were raised about the 'somewhat awful undertaking' of attempting to compete with earlier host venues over the standards of cuisine and entertainment.[71] The bar had been set high in Dublin three years earlier when levels of feasting attracted almost as much press comment as the scientific business of the congress.[72] It was this sense of competition with other host towns rather than the expectations of the visiting associations which created this pressure to spend. In fact, officials acting on behalf of the associations sometimes commented disapprovingly on the expenditure. While praising the success of the 'mighty meeting' at Newcastle, John Phillips noted that, 'The cost was enormous (£3,200 raised), not lavish (I assert), but yet such as to alarm small towns and make them measure their rooms, before inviting the philosophers'.[73] Levels of spending worried other members of the Association too. Roderick Murchison reported disapprovingly to Harcourt that £500 had been paid out to the Newcastle organisers: 'It is contended that when we go to these commercial towns they have not places to receive us and therefore are put to great expense, but I contend that no town is entitled to ask us who cannot house us and it is a bad example and precedent to defray any charges of this sort which may be made'.[74] At Glasgow in 1840, Murchison was horrified to learn the sum of money paid out by the local committee for use of the town theatre: 'they saddled the town with £800!!! For three nights of a theatre!! I expostulated but was too late. We have now resolved draw out a code of instructions for locals and ... shall insist on no such expenses and displays'.[75] The code was duly drawn up at a council meeting in December. It attempted to define the share of meeting expenses payable by the association and by the meeting venue, decreeing that, 'circulars and advertisements, registration of members, issue of tickets, printing lists of members and their addresses, printing daily sectional notices, post office department' were to be funded by the association, while 'provision of rooms,

---

70 *Proceedings and Reports of the Town Council of the Borough of Newcastle for 1838* (Newcastle, 1838), Quarterly meeting, Wed. May 1838.

71 G.T. Fox, quoted in the *Newcastle Courant*, 15 September 1837.

72 See for example, *The Times*, 19 August 1835; 4 September 1835.

73 John Phillips to William Harcourt, 17 September 1838, quoted in Morrell and Thackray (eds), *Gentlemen of Science. Early Correspondence*, p. 288.

74 Roderick Murchison to William Harcourt, 21 February 1839, quoted in Morrell and Thackray (eds.), *Gentlemen of Science. Early Correspondence of the British Association for the Advancement of Science*, p. 301.

75 Roderick Murchison to William Whewell, 29 September 1840, in Morrell and Thackray (eds), *Gentlemen of Science. Early Correspondence*, p. 341.

attendance, messengers and police' were the responsibility of the host town.[76] Even with this general guidance, local organisers sometimes sought more detailed information about expenditure. John Phillips of the BAAS sent the following statement of costs from the Dublin meeting in 1857 to Professor Fuller of Aberdeen ahead of its BAAS meeting in 1859:

Stationery	157	-	11	-	9	
Engraving and Maps	57	-	0	-	0	(map on card)
Printing	63	-	14	-	9	
Postage	19	-	18	-	8	
Advertising	74	-	1	-	9	(the association also advertise)
Office expenses – clerks	133	-	0	-	9	(lodgings clerk, post office secretaries)
Messengers and porters	85	-	7	-	6	
Refreshments	269	-	11	-	1	(this is about an average for a full meeting)
Concert subscription	60	-	0	-	0	
Hire of furniture	59	-	5	-	0	(this is sometimes replaced by a heavy carpenter's bill for fittings in the section and other rooms)
Sundries	62	-	10	-	8[77]	

By defining the costs to be met by the hosting locality in such a clear way, the BAAS was demarcating its own financial responsibilities and the purposes to which its own resources could be employed. Other associations adopted similar practices and ensured that monies raised from the annual fees paid by members and associates could not be used to settle debts incurred by the locality during the meeting because these were channelled back to the central coffers of the associations to cover their regular expenses such as the cost of publishing proceedings. Any attempt to call upon the central funds of the associations was likely to cause dispute, as in Edinburgh in 1863 when the local committee made a charge of £200 to the NAPSS to reimburse one of its members for his services. The ensuing row dragged on for several months and was widely reported in the Scottish press, and although the money was paid, it made the association anxious to be more explicit about the financial responsibilities of the host town at the stage of hearing future deputations.[77] Instead, to meet the costs of hosting a scientific meeting, towns had to fall back on their own resources. They had three principal means at their disposal. The first was the allocation of money from corporation funds. This could take a number of forms. Richard Rodger has shown how this could be achieved in Edinburgh via the 'Common Good' fund which the town council used for all manner of events and visits requiring a

[76] Bodleian Library, BAAS Dep.17, 2 December 1840.

[77] UCL Special Collections, Brougham MSS, 13,130, George W. Hastings to Lord Brougham, 22 December 1863.

measure of civic hospitality.[78] Elsewhere, town councils could agree to pay money from their own funds to help cover expenses. In Bristol, in the period immediately following municipal reform, some sleight of hand was needed to ensure that this proceeded smoothly. The decision of the old corporation to vote a sum of £200 from its funds for the purposes of the meeting was not permissible 'under the new order of things' in August 1836, so town councillors subscribed individually to the local fund to the same value as the original grant.[79] In Newcastle in 1838, a sum of £200 was unanimously voted at a quarterly council meeting towards the costs of room decoration and the entertainment and accommodation of foreign scientists at the forthcoming BAAS meeting.[80] At Lincoln in 1848, the corporation agreed to place a sum of £100 in the hands of the mayor and the local organising committee for the purposes of entertaining Royal Archaeological Institute members during their forthcoming visit to the town.[81]

The second fund of resources which towns could draw upon was the goodwill of local institutions and businesses, many of whom came forward to offer their services cut price or gratis. At Bristol in 1836, the local committee organising the British Association meeting offered the owners of the town theatre £140 for its use, but managed to obtain agreement from the gas company to light the building for free.[82] At Liverpool in 1858, a working men's instrumental band performed for free at the formal banquet, receiving a simple vote of thanks for its efforts from the local organising committee.[83] In Manchester the committee of the Royal Museum and Library in Peel Park, Salford, made their premises freely available to the British Association for its meeting in 1861,[84] and when the BAAS visited York in 1844, it was claimed that, 'Nearly all the edifices, institutions, establishments and exhibitions of interest in the city have been thrown open for the gratuitous admission of the members'.[85] In addition, local gentry in many localities could be relied upon to open up their houses and grounds and lay on refreshments for guests. The Royal Archaeological Institute delegates at Hereford were entertained by Lord

---

78 Rodger, 'The "common good" and civic promotion: Edinburgh', pp. 144–77.

79 *York Herald and General Advertiser*, 6 August 1836.

80 *Proceedings and Reports of the Town Council of the Borough of Newcastle for 1838* (Newcastle, 1838), quarterly meeting, 2 May 1838, pp. 16–17.

81 Neville Birch, 'The "archaeologians" visit Lincoln', in Christopher Sturman, (ed.), Lincolnshire *People and Places: essays in memory of Terence R. Leach (1937–1994)* (Lincoln, 1996), p. 173.

82 BRO 32079 (39) Minute Book of the Reception Committee for the British Association visit, 6 June and 20 June 1836.

83 NML MAL, John Towne Danson D/D/III/17/5a, Newspaper Cutting Book, *Northern Times*, 15 November 1858.

84 Bodleian Library, Dep. BAAS 59, Extract from the minutes of the proceedings of the committee of the Royal Museum and Library, Peel Park, Salford, held on the 10th day of September 1859.

85 *Leeds Mercury*, 28 September 1844.

and Lady Cornewall at Moccas Court in August 1877, their house 'beautifully situated on the banks of the Wye'.[86] Such contributions helped spread the burden of entertaining and catering for guests, and enabled local organising committees to offer some diversity of experience to visitors during meeting week.

The third, and probably the most important source of finance for the running of annual meetings took the form of monetary donations from supporters in the locality who were willing to respond to funding appeals. Most local organising committees launched subscription lists well in advance of the congress to ensure that sufficient funds were raised to cover their costs, but contributions were not always readily forthcoming. In Nottingham in 1866, local organisers preparing for the BAAS meeting reported that '2,200*l.* had been subscribed in Nottingham without any public appeal, but 4,000*l.* would be required. The expenses were incurred in fitting up rooms, in providing for "distinguished foreigners", in finding suppers and wines, printing *&c.*, *&c*'.[87] In Bradford, it was estimated that a public subscription of £4,000 would be needed to cover all of the expenses of the 1873 British Association visit because the lack of sufficient meeting rooms in the town meant that a temporary wooden structure would have to be constructed to provide additional accommodation.[88] Such appeals often received generous support but a downturn in trade in the hosting locality could make fundraising for the meeting a challenging proposition. Manchester withdrew an invitation to the NAPSS in 1864 due to 'the distress and want of money prevailing there'.[89] Unlike towns with more diverse economic profiles, its dependence upon cotton manufacture made it susceptible to slumps in trade,[90] and the shadow of the cotton famine still hung over the town in 1866 when it was feared that 'very slow progress was being made with the guarantee fund' ahead of the Social Science meeting in the city that year. Committee members found themselves having to solicit contributions by contacting individuals who had given financial support to previous congresses. In particular they decided to trawl through 'the list of the guarantors to the old British Association, and pick out such names as the members could influence'.[91]

Explicit criticism from within host towns of the costs incurred in hosting annual congresses rarely found its way into the provincial press, but rumblings of discontent were sometimes present. At the Lincoln corporation meeting,

86 *The Times*, 15 August 1877.

87 *Derby Mercury*, 3 January 1866.

88 *Leeds Mercury*, 21 November 1872.

89 UCL Special Collections, Brougham MSS, 13,143, George W. Hastings to Lord Brougham, 5 December 1864.

90 See Derek J. Oddy, 'Urban Famine in Nineteenth-Century Britain. The effect of the Lancashire Cotton Famine on Working-Class Diet and Health', *Economic History Review*, new series, 36/1 (1983): 68–86.

91 MCL, BR F306 S3, *Manchester Guardian*, 17 August 1866; *Manchester Courier*, 27 July 1866.

where the expense of entertaining the visiting archaeologists was discussed, one councillor alleged that it was illegal for town funds to be applied for such a purpose, and suggested that the mayor should foot the bill at his own personal expense.[92] Richard Davidson, addressing a public meeting at the Corn Exchange in Belfast to solicit donations to the local fund for the 1852 British Association meeting, could 'see one or two gentlemen shaking their heads at me' as he urged local merchants to open their cheque-books.[93] Similar feelings of disapproval can be inferred from the defensive stance adopted by the Manchester local organisers who, in preparing for the NAPSS meeting in 1866, found themselves having to counter 'a great misunderstanding on the part of many of the public respecting the purpose for which the guarantee fund was required' and to remind people that they 'did not touch a farthing' of the money for themselves.[94]

Perhaps because of these sorts of concerns, there were examples of relatively modest budgeting, such as in Manchester in 1879 where the local organising committee preparing for the meeting of the NAPSS estimated the expenses at £1,000, and Aberdeen where it was said that the NAPSS meeting in 1877 was put on for just £750,[95] but these were not the norm. Extra initiatives always had the potential to push costs up and the eagerness of towns to put on a more impressive show of hospitality and organisation than other venues meant that the temptation to stretch the budget was always strong. Clearly, where town pride was at stake, organisers were prepared to shoulder much greater expense in order to impress. One innovation introduced at early meetings of the British Association, and regularly adopted by host towns thereafter, was the production of special volumes, maps and other publications for distribution, often free, to meeting delegates. Town publishing firms were pressed into service to produce commemorative histories of the host town and its institutions or catalogues of special exhibits assembled for display during the meeting. At Bristol in 1836, each member of the British Association in attendance at the annual meeting collected, along with his ticket, a free memoir of the Bristol Institution and some 2,000 copies of an edition of the catalogue of exhibitions and places of interest open to members during the meeting were also produced.[96] In Southampton in 1846, a descriptive account of the town and its vicinity with a geological guide to the Isle of Wight and its scenery was produced by the local committee, 'with great care, and at considerable expense'.[97] In Exeter, a map was drawn up by the

---

[92] Birch, 'The "archaeologians" visit Lincoln', p. 173.

[93] *Belfast Newsletter*, 16 April 1852.

[94] MCL BR F306 S3, *Manchester Guardian*, 17 August 1866; *Manchester Courier*, 17 August 1866.

[95] MCL BR F306 F2, *Manchester Guardian*, 22 March 1879; *Manchester Guardian*, 29 May 1879.

[96] BRO, 32079(39) Minute Book of the Reception Committee for the British Association visit, 15 August and 19 August 1836.

[97] *Daily News*, 8 September 1846.

local organising committee for the 1869 British Association meeting, showing the location of all manner of geological features and places of interest throughout the south west of England, from rock outcrops and fossilised remains to mines, museums and castles.[98] In Bradford in 1873, a town guide was distributed to BAAS members. One visiting newspaper correspondent noted that 'The people of Bradford are doing their best certainly to tell visitors all that is to be told of their town. A shilling Guide-book is politely offered gratis to each stranger, and the local committee present to the members, associates and ladies, a most interesting essay on Bradford and its worsted manufacture, by Mr Taylor'.[99] The aim was to draw their attention to the attractions of the place and advances in commerce and manufacture, but it was a costly gesture, with local organising committees, or rather subscribers to their guarantee funds, usually footing the bill.

Alternatively, newspaper columns sometimes served a similar function, printing town histories for the benefit of visiting societies. In July 1868 the *Leicester Chronicle* devoted generous column space to a 'Historical Summary' of the town for visitors attending the RASE show, who 'may be desirous of knowing something of the town in days bygone'. The account highlighted significant events in the urban development of Leicester, from the granting of a fair in 1305 to the town charter of 1487 and the visit of Charles I in 1642.[100] Such chronological surveys were designed to help outsiders make sense of the town and its role in national life and, all importantly, to convey a sense of its progress and improvement. The *Chronicle* editors omitted events of 'merely local interest' but proceeded to detail 'the extension of education, the progress of commerce and the growth of knowledge' as subjects more suitable for 'the enquiring visitor'. The production of street maps with key institutions highlighted performed a similar function, albeit in a more visual form. Organisers of the Bristol BAAS meeting in 1836 followed the example set by their predecessors at Edinburgh and produced a small card printed with a map of the meeting locations.[101] Once established, the traditional meeting map became used not just as a guide for visitors to find their way to the meeting rooms, but as a vehicle to exhibit the progress and enlightenment in the town through the presence of key urban institutions. The map produced for the 1866 NAPSS meeting in Manchester plotted the locations of the infirmary, the children's hospital and the house of industry, as well as the major hotels, railway stations, the museum, library and Royal Institution.[102] Although these key buildings were rather dispersed around the streetscape, as the Piccadilly district had given way to new focal points for

---

98 Westcountry Studies Library [WSL], WES/1869 D'Urban, Map edited by W.W.M. D'Urban for the local committee of the BAAS Exeter Meeting, August 1869.

99 *Daily News*, 18 September 1873.

100 *Leicester Chronicle*, 11 July 1868.

101 BRO, 32079 (39) Minute Book of the Reception Committee for the British Association visit, 10 May 1836.

102 MCL, F306 S3, Social Science Association, Manchester Meeting 1866.

the wealthy middle classes, such as St Anne's Square,[103] the principal streets and squares were named and the presentation of an orderly overview of the town must have been reassuring for newcomers expecting to find a labyrinthine industrial maze. Once armed with these maps the expectation was that, weather permitting, delegates would spend time walking the streets and viewing the key features of the urban landscape. At Glasgow in 1874, NAPSS delegates enjoyed weather 'of a very pleasant character', thus enabling them, 'in the intervals between Congress engagements ... to see the city under the most favourable conditions'.[104]

If the printing of maps and complimentary guides for visitors had the potential to inflate expenditure, so too did the supply of hospitality at annual meetings. Catering was often the barometer by which outsiders judged the success of the hosts but it was a difficult element to get right: too lavish and participants might be accused of putting the interests of gastronomy before those of science; too plain and organisers appeared mean and risked unfavourable comparison with other venues. At Dundee in 1867, the corporation pulled out all the stops with its menu for a banquet held in the Albert Institute Hall on the 11 September where British Association guests dined on a sumptuous array of turtle and hare soups, salmon, game, roast beef, ham and tongues, plum puddings, tarts and jellies, desserts and ices.[105] At the annual Royal Agricultural Society shows, the issue of dining was particularly problematic. Different types of catering had to be provided at these large-scale events, the precedent for which was set at Liverpool in 1841 where, on the first day of the show, the council of the Association dined at Lucas's Horse Repository with Mr Pusey, MP, in the chair. Later in the week a grand dinner of 3,000 people was held in a specially constructed pavilion.[106] The tradition of having a 'council' dinner and a 'pavilion dinner' continued but standards of food and service were an ongoing subject of debate in RASE circles. At Derby in 1843, the council dinner was laid on by Mr Neale of the town's County Tavern. While the *Derby Mercury* claimed that the standard of the dinner 'did him great credit', RASE council members were less impressed and summoned him to attend a meeting to explain himself. He claimed that in the crush of attendees, waiters had been unable to reach the tables and people had begun to eat before they got into the room. The Council resolved that he had 'not fulfilled the terms of his contract for the Council Dinner'.[107] At Norwich in 1849, RASE members considered both dinners to have been poor and the prospect of dropping one of them was mooted by its council members soon afterwards. It was thought better

---

103 See James R. Moore, 'Urban space and civic identity in Manchester, 1780–1914: Piccadilly Square and the art gallery question', *Transactions of the Historic Society of Lancashire and Cheshire*, 153 (2004): 87–123.

104 *Glasgow Herald*, 1 October 1874.

105 Bodleian Library, BAAS Dep. 150, Printed material and correspondence relating to the Dundee meeting, 1867, menu card.

106 *The Times*, 22 July 1841.

107 MERL, SR RASE/B1/2 Minutes of Council, 1842–1844 (rough copy), 2 August 1843.

to have one good dinner than two bad ones, but more effort had to be made to make the dinner interesting and provide good speeches, because people 'want something more than cold dinner and bad wine'.[108] When the pavilion dinner was dropped for the first time at Salisbury in 1857, however, the Society came in for heavy criticism. A single council dinner was held at a pound a head which had 'an air of exclusiveness which is little relished by the tenant farmers' who, as a result, organised their own dinner the following evening.[109]

Organising towns were anxious at all costs to avoid the prospect of being blamed for a poorly organised or attended dinner, or for failure to cater sufficiently for the needs of those attending the meeting. The mayor of Canterbury challenged RASE council members who visited that town ahead of the annual show there in 1860 and suggested that the dinner should be locally managed. He 'objected to a dinner taking place under the auspices of the society while the arrangement, cost, responsibility and patronage should be left to local authorities'.[110] It was a question of control which revealed the degree to which towns regarded themselves as active partners in the staging of these annual visitations with the right to negotiate over how things should be done. When it came to evening arrangements, especially, the visitors were largely in the hands of local organisers. At Glasgow, for example, Hastings and Brougham had to re-arrange their plans to accommodate a private dinner given by the Lord Provost in the picture gallery at the corporation at which Brougham's presence was expected. Hastings advised Brougham that 'it will be as well to acknowledge the courtesy of the Provost, who in common with the rest of the Corporation has done much for the meeting. There will be no speeches and you and get away at 7'.[111]

Dinners such as this were just one element of the activities which took place outside the sectional rooms and lecture theatres during meeting week. For attendees, it was important to have opportunities to meet and mingle with fellow enthusiasts and to have some respite from intellectual business of the congress. As John Robison wrote to Phillips from Edinburgh in 1834, 'The exertion of both body and mind which is required to sustain the state of permanent activity during 12 hours per diem throughout a whole week, is too great ... relaxation would be generally relished'.[112] But different views prevailed as to how best to use the available time, and visiting associations found themselves having to negotiate with host towns which had their own agendas. Local organisers viewed evening events as an opportunity for place

---

108 MERL, SR RASE/B1/4, Minutes of Council, 1848–1853 (rough copy), 1 May 1850.

109 *The Standard*, 23 July 1857.

110 MERL, SR RASE B1/6 Minutes of Council, 1858–1868 (rough copy), 7 March 1860.

111 UCL Special Collections, B MSS, 7660, George W. Hastings to Lord Brougham, 12 September 1860.

112 John Robison to John Phillips, 5 October 1834, quoted in Morrell and Thackray (eds), *Gentlemen of Science. Early Correspondence*, p. 193. Robison was the son of an Edinburgh professor of natural philosophy who became active in the Mechanical Science Section of the BAAS. See Morrell and Thackray, *Gentlemen of Science. Early Correspondence*, p. 37.

promotion. Whether through the use of landmark buildings for the staging of dinners and balls, or the showcasing of talent from the region in exhibitions or concerts, host towns took advantage of the presence of large numbers of visitors in need of recreation to display their best assets. The 1859 NAPSS meeting at Bradford proved to be a public stage for one of the town's great cultural institutions, the Bradford Festival Choral Society, which entertained delegates at an evening gathering with an excerpt from their repertoire and a paper by one of their founder members, Samuel Smith, on the organisation and constitution of the society.[113] At Dundee the Philharmonic and St Cecilia Society gave a complimentary concert in the Kinnaird Hall during the BAAS meeting, featuring excerpts from Mendelssohn's *Elijah* and works by Handel, Mozart and Rossini.[114] At Leeds in 1858, the town's Photographic Society put on an exhibition of photographs at the town hall. At a meeting where a number of papers on photography were delivered in the Chemical Science section, it was an opportunity for the members of the local society to display their prowess to visiting experts.[115]

Entrepreneurs and businesses independent of the local organising committee also saw annual meetings as a commercial opportunity and peppered the week with additional attractions. At Cork, the proprietor of the local theatre announced that 'Miss Novello and her sister, Miss Sabilla Novello' had been engaged to perform during the week of the BAAS meeting in 1843.[116] RASE shows, in particular, were magnets for local entertainers and salesmen to ply their trades. At Derby in 1843 it was reported that:

> The whole neighbourhood of the large Pavilion was crowded with suttling booths, bazaars, eight divans, shows, moveable theatres, stalls for games of chance, swings, and all the paraphernalia of a country wake or fare ... There were wild beasts and tame beasts, jugglers, pyrotechnists ... to say nothing of the arboretum, the museum, the theatre, and other places of spectacle and amusement native to the town.[117]

113 NML MAL, John Towne Danson D/D/III/15/3, Printed Programme of the Bradford Meeting, 1859.

114 Bodleian, Dep. BAAS 150, Printed material and correspondence relating to the Dundee meeting, 1867.

115 Bodleian, Dep. BAAS 146, Printed material for annual meetings, 1831–59, Leeds 1858. For details of papers delivered on photography see *Leeds Mercury*, 28 September 1858.

116 *The Times*, 18 August 1843. Clara Anastasia Novello was a renowned singer of opera and oratorio. See Victoria L. Cooper, 'Novello, Clara Anastasia (1818–1908)', *Oxford Dictionary of National Biography* (Oxford, 2004), accessed 15 December 2012.

117 *Derby Mercury*, 19 July 1843. A suttling booth was a place where food and drink were sold and divans were possibly smoking rooms, see 'suttle, v'. and 'divan, n'. *Oxford English Dictionary* (online edition, June 2011), accessed 22 August 2011.

At Lincoln in 1854, the RASE show was accompanied by a similar plethora of attractions including Wombwell's Menagerie and a variety of drinking and dancing booths.[118] There were no comparable sights at British Association meetings, but abundant other sorts of entertainments were laid on. At Exeter in 1869, local organisers issued cut-price tourist tickets to members and, when it emerged that the manager of the new Exmouth Imperial Hotel was issuing free railway tickets between Exeter and Exmouth, they encouraged 'purveyors of entertainment in the city' not to be 'outdone in generosity' by the latter town.[119]

In short, host towns filled the time available outside of scheduled lectures and meetings with a bewildering and sometimes exhausting array of entertainments. John Phillips of the BAAS came to regard the evenings of meeting week with a sense of dread, and wrote to Harcourt from Bristol in 1836: 'As to the evening meetings, we have a great embarrassment and must not yet decide what is to be done on some evenings. I mean whether to grow hot in the Theatre, or cool in the gardens of the Horticultural and Zoological Society …'[120] Others expressed concern about the potential of entertainments such as these to distract attention from the scientific business of the meetings: William Pengelly, writing to his wife from the British Association meeting at Exeter in 1869, commented that in some of the sectional meetings, 'The audience was small as there were the more popular attractions of a Bazaar and Flower Show'.[121] RASE members, too, feared for the reputation of their Society if amusements gained the upper hand over agricultural improvement work at the annual show. These concerns came to a head at Manchester in 1869 when, for the first time, a horse-jumping competition was staged at the showground. The event showed the potential of show-goers to by-pass the core agricultural business of the show, diverting the usual crowds from attending the implement trials.[122] Even where lectures were laid on in the evenings, these were often designed to attract a large local audience and sometimes lacked intellectual rigour. As BAAS secretary John Phillips put it, evening lectures could 'present great difficulties, it being not easy to make them at once popular and respectably scientific'.[123]

Organisers in the localities were not oblivious to these concerns and made concerted attempts in their programme of official entertainments to marry recreation with points of educational interest. The conversazione provided them with the perfect vehicle. An evening gathering combining lecture, exhibition, discussion, musical entertainment and refreshment, conversaziones were one of the most popular events used by clubs and societies, large and small, to provide

---

118 *The Morning Post*, 19 July 1854.

119 Bodleian Library, Dep. BAAS 152: Printed material and correspondence relating to the Exeter meeting in 1869.

120 John Phillips to William Harcourt, 8 August 1836, quoted in Morrell and Thackray (eds), *Gentlemen of Science. Early Correspondence*, p. 235.

121 Quoted in, Pengelly (ed.) *A Memoir of William Pengelly of Torquay, FRS.*, p. 130.

122 *Sheffield and Rotherham Independent*, 16 July 1869; *Daily News*, 21 July 1869.

123 Bodleian Library, Dep BAAS 132, Notebook of John Phillips, 1834, p .9.

their members with an opportunity to meet, converse and engage with speakers and exhibitors on all manner of topics.[124] The typical scene at a conversazione saw a polite and fashionable crowd juxtaposed with an assortment of objects and instruments. At annual meetings these events seemed to satisfy delegates' needs for some light, but informative, relief from the business of the lecture halls and discussion rooms. At the Social Science Association's meeting at Manchester in 1879, an evening conversazione was hosted in the town hall by the mayor and corporation. Along with musical entertainment provided by an organist and a military band, attendees were treated to a demonstration of 'one of Edison's loud-speaking telephones ... communicating with an apartment at the opposite end of the building'. For inventors like Edison, events such as this provided important opportunities to exhibit and demonstrate new instruments and discoveries to an admiring audience.[125] A more incongruous exhibit was 'a quantity of Cambridgeshire butter, in yard lengths, a notice being attached to the effect that the cows, which were stall fed, were milked and managed by boys in the Cambridge refuge under 15 years of age'.[126] Soirées featuring exhibitions of microscopes became almost routine at BAAS meetings in the late 1850s and early 1860s, a period when the much improved magnifying powers of achromatic microscopes were being applied to a wide range of scientific fields.[127] At the Manchester meeting in 1861 a 'Microscopic Soirée' was held at the Free Trade Hall where, in the midst of 'a brilliant assemblage of upwards of 2,500 members of the Association ... were displayed about 50 microscopes, of various size and powers' which had been loaned for the occasion by the gentlemen of the neighbourhood. On another evening of the same congress, a 'Natural History Soirée' was held, featuring an exhibition of botanical and zoological specimens contributed by members of the Manchester Field Naturalists Association.[128]

Events such as these provided the town with a means of exhibiting some of its own scientific riches to the learned visitors and demonstrated that knowledge exchange was not necessarily regarded as a one-way process during meeting week. In fact, in the running of exhibitions, excursions and conversaziones, local organisers were treading familiar territory. These kinds of events were the mainstay of local scientific and learned societies throughout the provinces and there was thus a wealth of organisational expertise on which local reception

[124] Samuel J.M.M. Alberti, 'Conversaziones and the Experience of Science in Victorian England', *Journal of Victorian Culture*, 8/2 (2003): 208–30.

[125] Iwan R. Morus, 'Manufacturing Nature: science, technology and Victorian consumer culture', *British Journal for the History of Science*, 29/4 (1996): 425.

[126] MCL, BR F306 F2, *Manchester Guardian*, 2 October 1879.

[127] Robert H. Nuttall, 'The Achromatic Microscope in the History of Nineteenth-Century Science', *The Philosophical Journal*, 11 (1974): 72.

[128] Bodleian Library, Dep. BAAS 147, Printed material for the annual meetings, 1860–63, 'Proceedings of the thirty-first meeting of the British Association for the Advancement of Science', reprinted from the *Manchester Guardian*.

committees could draw.[129] At Lincoln in 1848, for example, it was no accident that the local organising group for the Royal Archaeological Institute meeting that year included the architect and antiquarian E.J. Willson and Captain Arthur Trollope of Casewick, who had his own valuable collection of coins. At the opening meeting a vote of thanks was given for the Spalding Gentlemen's Society, which was credited with being 'of very great service to the county antiquary', and later in the week Willson presented a paper on 'The old Archiepiscopal Palace at Lincoln'.[130]

This capacity for the host venue to 'teach' as well as to 'learn' from visiting associations extended well beyond the confines of the evening soirées and conversaziones. In the case of Royal Archaeological Institute meetings, the town and its immediate locality provided the subject matter for study by visiting delegates. Viewings of archaeological remains, architectural attractions, antiquarian artefacts and sites of historic interest in the locality formed the mainstay of RAI meetings, with papers delivered to audiences in lecture rooms a lesser feature. At Leicester in 1870, the range of sites for delegates to inspect included 'the Norman hall of its ancient castle, the "newarke" of the castle, numerous Roman pavements … and the noted mass of Roman masonry called the "Jewry Wall"'.[131] Other, more portable, objects were assembled in temporary museums which were a feature of RAI meetings. Here local gentry and clergy loaned objects from their collections for display to assembled visitors. At Newcastle upon Tyne in 1852, the temporary museum was declared 'the most remarkable feature of the congress'. Its exhibits included Celtic arrow heads loaned by Mr John Ord of Darlington; part of a jet necklace, from the collection of the Newcastle Society of Antiquaries; and a bronze spear head, found near Barnard Castle, Durham.[132] As well as offering an outlet for the treasures of local collectors and a shop window for the archaeological riches of the host town and its region, the assembling of artefacts in these temporary displays was an important signifier of the scientific value of archaeology and a feature which distinguished it from antiquarianism.[133]

In the business of the sectional meetings, too, there was scope for the presiding town to influence the content of scientific discussion. The BAAS' sections were viewed, by participating scientists, as the core activity of the congress. Murchison wrote that, 'I look upon the morning meetings as the very essence of the Association, for although our reports constitute our real claims in the literature of science, the sections are the chronometers which mark the rate of the current of living enterprise and in short give rise to all our recommendations, reports and so

129 See for example the discussion of conversaziones and exhibitions in Diarmid A. Finnegan, 'Natural History Societies in late Victorian Scotland and the Pursuit of Civic Science', *British Journal for the History of Science*, 38/1 (2005): 53–72.

130 *Daily News*, 27 July 1848.

131 *The Times*, 4 July 1870.

132 *Morning Post*, 28 August 1852.

133 Ebbatson, thesis, p. 23.

Figure 3.4 Temporary Museum for the Archaeological Institute meeting at Edinburgh in 1856, *Illustrated Times*, 9 August 1856 (author's collection)

forth'.[134] Although the selection of papers for delivery in the sections was a matter for the central council rather than the local organising committee in the case of both the BAAS and the NAPSS, suggestions for appropriate local topics were both offered to, and solicited by, association representatives who liaised with local organisers during the months of planning. Once the BAAS' intention to hold its 1853 meeting in Hull had been announced, the editors of the local newspaper made a number of suggestions for appropriate subjects for study: 'The great delta of the Humber has never been thoroughly examined; the cliffs from Spurn Point to Flamborough Head will be objects of interesting geological survey; and to the archaeologist, Beverley Minster, Thornton College, and our own venerable parish church, will be objects of no mean regard'.[135] At the Social Science congress in Norwich in 1873, the meeting president, Lord Houghton, commented that 'the Association would be glad if the inhabitants of the town and neighbourhood would participate in the sectional proceedings, bringing out something of the peculiar characteristics of the East of England'.[136] Hastings, on behalf of the NAPSS, often gave local committees a very precise steer in the direction of local subjects that would be appropriate for inclusion. He explained to organisers in Liverpool in 1858 that while the central committee retained control of all arrangements relating to papers, 'with respect to the local papers, what we shall be very glad to obtain would be accurate statements of the actual condition of different classes of the population in Liverpool – as to the employment of the population – as to their

134 Roderick Murchison to William Harcourt, 18 July 1838, quoted in Morrell and Thackray, *Gentlemen of Science. Early Correspondence of the British Association for the Advancement of Science*, p. 266.

135 *Hull Packet*, 10 September 1852.

136 *Leeds Mercury*, 2 October 1873.

sanitary condition – as to the amount of crime – and in short the general social and economical state of the different classes of this very large town'.[137]

The result was that contributions focusing on aspects of the host town and its locality featured in the proceedings of many of the different sections. At Hull in 1853, William Lawton presented a paper on 'The Meteorology of Hull' to the Mathematics and Physics section; J.D. Sollitt addressed the Chemistry section on 'The Chemical Constitution of the Humber Deposits'; Charles Beckett presented a paper to the Geography and Ethnology section on 'Dialects North and South of the Humber'; and in the Statistics section, papers were presented by Dr Henry Cooper on 'The mortality of Hull in the Autumn of 1849' and 'The Prevalence of Diseases in Hull'.[138] During the Social Science congress in Liverpool in 1858, papers were contributed by Dr Hume who addressed the Education section on 'The Education of the Poor in Liverpool with some suggestions for its Improvement'.[139] In the Public Health section, Mr James Newlands C.E., borough engineer, read a paper on 'The Past and Present Sanitary Condition of Liverpool' and in the Social Economy section a paper on 'The Objects and History of the Liverpool Co-operative Provident Association' was contributed by the president of that association, Mr John Wilson.[140] According to the *Daily News'* correspondent, the British Association's programme at Bradford in 1873 was heavily influenced by its meeting in a town which was 'essentially the modern product of mechanical invention and science'. The local organising committee commissioned an essay on Bradford's worsted manufacture for distribution to members and the Chemistry Section was described as 'a particularly strong department at this year's meeting, because of the important part chemistry plays in the woollen industries'.[141]

There was scope, too, for examining some of the interesting features of the meeting's locality during the excursions which formed an important part of the programme of BAAS and NAPSS meetings. At Newcastle upon Tyne in 1838, the British Association's geologists examined coastal features around Tynemouth. A party of around 200 made the trip out along the Northumberland coast and listened to Adam Sedgwick's explanations of the various phenomena from vantage points on the beach and cliffs.[142] Sedgwick later recounted the trip in a letter to his wife who, 'would have been amused at the picturesque group clustering among the rugged precipices of a noble sea-cliff or congregating on the sand below, while I addressed them at the utmost stretch of my voice six different times, from some

137 NML MAL, John Towne Danson Collection D/D/III/15/3, Proceedings of the meeting of the General Committee of the NAPSS, 5 May 1858.

138 *Report of the Twenty-Third Meeting of the British Association for the Advancement of Science held at Hull in September 1853* (London, 1854), pp. vi–x.

139 *Liverpool Mercury*, 15 October 1858.

140 *Liverpool Mercury*, 16 October 1858.

141 *Daily News*, 18 September 1873.

142 *Newcastle Courant*, 31 August 1838.

projecting ledge ...'[143] In Manchester, visiting social scientists attending the annual congress of 1866 were taken to see examples of 'the great co-operative movements with which Manchester had had to do', following a suggestion from Hastings that this would be an appropriate subject of interest.[144] The local organising committee arranged an excursion to Sabden Bridge and Rochdale where delegates could view examples of institutions run on the co-operative principle. The opening of a new cotton mill at Whalley, near Sabden Bridge, was one of the day's main attractions, an enterprise which speakers praised for innovatively applying the co-operative principle to industrial production.[145]

Excursion destinations such as these not only satisfied the social scientists' appetite for studying the distinctive features of the region, but also gave local organisers an opportunity to show off a specialist feature or project. Local organisers of the British Association meeting at Bristol in 1836 milked this approach to its full potential when they arranged for members of the BAAS to attend the foundation stone-laying ceremony for the Clifton Suspension Bridge early on the Saturday morning of the annual meeting week. The stone-laying ceremony was performed by the BAAS president, the Marquis of Northampton, flanked by I.K. Brunel, chief engineer, and representatives of the trustees of the scheme. The whole event was staged for maximum visual effect with the Marquis arriving 'in a carriage drawn by six greys, the postillions wearing crimson satin jackets and black caps with gold tassles'. The following procession included a band of music, a group of flag bearers and the architect carrying the foundation plate, the silver trowel and the mallet, and the whole procedure was reputedly watched from the banks of the river by 'tens of thousands of the inhabitants of the city and neighbourhood' as well as by spectators aboard vessels on the river, which were specially decked out with streamers and flags for the occasion.[146] The showcasing of such an ambitious engineering feat was perfectly timed to impress the scientific visitors to Bristol with an impression of improvement and progress and, at the celebratory breakfast held afterwards at the Gloucester Hotel, Clifton speakers, buoyed with optimism, predicted that the majority of the company present might return to see the opening of the bridge in three years' time.[147]

At RASE shows, council members as well as local organisers looked for opportunities to use the expertise of the host town and its locality to full advantage. At Cardiff in 1872, William Menelaus, manager at Dowlais Ironworks at nearby Merthyr Tydfil since 1856, was appointed one of the judges charged with the

143 Clark (ed.), *The Life and Letters of the Reverend Adam Sedgwick*, vol. 1, pp. 813–15.

144 MCL, BR F306 S3, *Manchester Guardian*, 1 March 1866.

145 See comments of Mr Morrison, MP for Plymouth, *Caledonian Mercury*, 9 October 1866.

146 *Bristol Mercury*, 3 September 1836.

147 The project was plagued with financial setbacks and was not completed until 1864. See A.E. Cottrell, *The History of Clifton Suspension Bridge* (6th edn., Bristol, 1956), p. 38.

testing of the portable steam engines on show.[148] A well-known industrialist locally, he also had wider connections in industry as a founder member of the Iron and Steel Institute of Great Britain, established in 1869,[149] and was therefore an ideal choice. It was acknowledged that different elements of the RASE show took on particular importance depending on the place of meeting. In 1871 the two great divisions of livestock and implements were assembled at Wolverhampton, 'But, as it is sitting in such a town ... the latter division will have additional importance attached to it' and, consequently, a silver cup was donated by Lord Vernon, the RASE president for the year, for the best steam-powered machine for soil cultivation.[150] At Carlisle, though, interest in the show of stock 'effectually drained the implement department of its visitors'.[151] The northerly location of the show attracted a number of Scottish breeds, notably Clydesdale horses and Galloway cattle which 'proved a successful novelty'.[152] In this way, the role of host towns extended well beyond the provision of a venue for learning to become participants and shapers in the exchange of knowledge.

The way in which annual meetings were run by the towns hosting the various parliaments of science in the mid-nineteenth century casts a revealing light on how these towns operated and what they saw as their most distinctive features worth promoting to the outside world. That they had the wherewithal to set aside ideological differences, to spend prodigiously and to add their own embellishments and attractions to the prescribed programme of lectures and discussions illustrates how valuable these congresses were perceived to be as status-enhancing tools. But in tailoring and enhancing the annual meeting in this way, local organisers were rarely doing anything new. Rather they were drawing on established traditions of civic ritual, dining etiquette and the wealth of experience in their own provincial scientific societies to stage events and socials which would provide meeting delegates with a suitable measure of hospitality while also focusing their attention on the assets of the locality. The annual meeting, though an external visitation, thus became an expression of civic identity as much as it was a festival of knowledge, in every urban setting in which it was held. As the parliaments of science continued their perambulations around provincial Britain over successive decades, it was the peculiarity of place which gave distinctiveness to each meeting and shaped the delegate experience.

---

148 *The Times*, 13 July 1872.

149 John A. Owen, *The History of the Dowlais Ironworks, 1759–1970* (Merthyr Tydfil, 1972), p. 68.

150 *Birmingham Daily Post*, 23 July 1871.

151 *Preston Guardian*, 28 July 1855.

152 *Lancaster Gazette and General Advertiser*, 28 July 1855.

Chapter 4
Experiencing the Meeting

Issues of knowledge reception and of audience engagement have featured prominently in studies of the history of science in recent decades. Much research has focused on the eighteenth century and the role of landed and professional classes in scientific meetings and lectures.[1] Levels of attendance at meetings, the length of discussions and the degree of participation in debate are all measures which historians have employed to assess audience responses to scientific ideas. However Steven Shapin's contention that, 'Consideration of the audience for science is pointless if it cannot be shown that the audience is active rather than passive, influential rather than submissive', has been questioned by research focused on the Victorian era, and specifically on meetings of the BAAS. [2] His approach privileges a definition of 'audience' as those present in person at a scientific meeting or event, who engaged with and responded to the ideas they saw and heard presented. Such a definition, while appropriate for the purpose of examining the communication of scientific ideas, is less applicable here. In this study, in which the focus ranges beyond the scientific content of parliaments of science to a consideration of them as urban events, a broader definition of audience is required. Similar to the way in which the study of eighteenth- and nineteenth-century elections has revealed levels of involvement among the non-voting public,[3] it will be suggested here that people beyond the confines of the meeting rooms and lecture theatres found ways of experiencing or engaging with the congresses taking place in their town. In pursuit of this, a much wider definition of participation is adopted, encompassing not just engagement with the knowledge content of the meetings, but evidence of interest in the excitement and spectacle generated by these occasions.

This chapter provides an assessment of the ways in which both the attendees and the wider populace of Britain's provincial towns encountered the annual meetings of the BAAS, NAPSS, RASE and the RAI. The audience at the scientific

---

[1] See for example, K.A. James, '"Humbly dedicated": Petiver and the audience for natural history in early eighteenth-century Britain', *Archives of Natural History*, 31/2 (2004): 318–29; Jan V. Golinski, 'Utility and audience in eighteenth-century chemistry: case studies of William Cullen and Joseph Priestly', *The British Journal for the History of Science*, 21/1 (1988): 1–31.

[2] Steven Shapin, 'The audience for science in eighteenth-century Edinburgh', *History of Science*, 12/2 (1974): 96; Higgitt and Withers, 'Science and Sociability', p. 3.

[3] Frank O'Gorman, 'Campaign Rituals and Ceremonies: the Social Meaning of Elections in England, 1780–1860', *Past and Present*, 135 (1992): 79–115.

talks and exhibitions of meeting week experienced these events at close quarters and details of their responses and interactions found their way into newspaper reports, transactions and other official accounts. But whenever the delegates and activities of the congress spilled out into the streets and neighbourhoods of the host locality, there was potential for a much wider cross-section of the public to engage with the parliaments of science. Some reports of these encounters filtered through into the press as well as into the occasional diary entry or letter penned by an attendee. These disparate sources enable the historian to sketch a picture, albeit an impressionistic one, of a diverse urban audience and a broad spectrum of participation in which levels of engagement were affected by a number of different factors. Peoples' ability to access the proceedings, the nature of their surroundings and the effectiveness with which information was communicated by speakers, as well their own level of expectation and interest in the various attractions on offer, all played a part in shaping the meeting experience. As far as organisers were concerned, meanwhile, the behaviour of the public both inside and outside the meeting room was the one key variable which no amount of careful preparation and planning could account for. Despite measures put in place to control the character and conduct of audiences, and despite sometimes fixed preconceptions about how different attendees would behave, especially women, there was a limit to which the response of the fee-paying delegate or the curious onlooker on the streets could be either predicted or influenced.

Host towns and their local organising committees, along with the leading figures in the learned associations, were required to perform a delicate balancing act when it came to setting the parameters for public involvement in the annual meeting. The view that the annual meeting could bring enlightenment, knowledge and learning to the populations of the towns and regions where they were staged motivated at least some of the activists in all four of the associations under scrutiny in this book.[4] The trouble was that the idea of spreading a message of knowledge and scientific enrichment to the masses was not a straightforward one. The question of how to generate widespread interest in scientific meetings, while at the same time preserving their intellectual credibility, vexed each of the four associations to a greater or lesser degree. It was a problem made more difficult by the fact that the distinction between 'popular' and 'scientific' was not always obvious in this period.[5] There was recognition, in the British Association for example, that the straightforward reading of a scientific paper, while it might be the simplest way to convey knowledge, would have little popular appeal. Less certain was what alternative methods of communicating ideas could be employed. Correspondence preceding the first British Association meeting at York, for instance, included a letter from William Swainson of St Albans to John Phillips in September 1831 warning that, 'Little interest I apprehend will be attached to the simple reading

[4] Although as we saw in Chapter 1 this was probably less important for the Royal Archaeological Institute than for the other three bodies.

[5] Kitteringham, thesis, p. 230.

of scientific papers', and a hope that, 'the meeting would have the appearance of a scientific fair', rather than 'of a grave, formal, dull assembly of learned men'.[6] Evening events, which had the potential to draw in the widest audiences, were often given over to more entertaining modes of activity, such as at Oxford in 1860 when 'A popular lecture, explanatory and descriptive on the lost polar expedition with large diagrams and sketches' was scheduled to be held in the town hall at four o'clock one afternoon, and evening soirées were organised for the following Monday.[7] But it was these very sorts of events which fuelled criticism of the Association by those who doubted its scientific credibility. Even by the 1860s, when *The Times* had come to accept the BAAS meeting as 'the great event of the scientific season', it continued to lament the extent to which 'the social element still encroaches on the scientific ... There must be soirées and conversaziones, and the associations derived from these will give a certain dilettante air to the more serious proceedings'.[8]

Despite the sceptical voices, the delegate numbers for the annual congresses of RASE, the BAAS and the NAPSS suggest some degree of success in appealing to popular interests. These were attractive events which large numbers of people were keen to attend. At its height in the 1860s, the NAPSS attracted in excess of 2,000 people to its week-long congresses. The 2,872 who attended the Glasgow meeting in 1860, all but matched by the 2,830 in Edinburgh three years later, were held up as examples for later host towns to aspire to.[9] These figures, which included members, associates and ladies with 'transferable tickets' attending the sections, evening meetings and opening and closing addresses, compared well with the numbers attracted to the meetings of the British Association. Its largest recorded attendances at nineteenth-century annual meetings were at Newcastle in 1863 with 3,335 and Manchester in 1887 with 3,838.[10] These figures, based on ticket sales, did not reflect the number who attended free events or caught snatches of open air lectures, and thus we can surmise that the actual number of participants at BAAS events would have been greater than the official statistics. Nevertheless, attendances at the annual shows of the Royal Agricultural Society of England were on an altogether different scale. Without the limitations of the size of meeting rooms and lecture halls, RASE was able to accommodate vast numbers at its cattle shows and implement exhibitions in fields and outdoor spaces located in or adjacent to the host town. This resulted in audiences running into the tens of thousands, peaking at 189,000 at the Manchester show in 1869.[11]

---

6 Bodleian Library, Dep BAAS 1, Correspondence of John Phillips (foundation volume), William Swainson to John Phillips, 15 September 1831.

7 Bodleian Library, Dep. BAAS 147, printed material for the annual meeting at Oxford.

8 *The Times*, 3 October 1862.

9 MCL BR F306 F2, cutting from *Manchester Courier*, 25 September 1879.

10 Howarth, *The British Association for the Advancement of Science. A Retrospect*, p. 118.

11 Goddard, *Harvests of Change*, p. 35.

But there was a clear tension between the desire for a large attendance at meetings and a concern to preserve intellectual credibility. The problem of how to strike this balance was a major preoccupation for the leading lights of the associations. It was an issue of particular resonance for the Royal Archaeological Institute whose meetings drew in small numbers in comparison with the other associations. The 500-strong gathering at a reception hosted by the Marchioness of Ripon during the 1874 RAI meeting was regarded as an exceptional turn out, 'probably the largest that has ever attended such a gathering'.[12] The figure was unremarkable by BAAS and NAPSS standards, but, in part, the smaller scale of the archaeologists' annual gatherings was a result of the Institute's deliberate policy, reflecting members' uncertainty about whether mass urban meetings were really effective vehicles for serious intellectual endeavour. As we saw in Chapter 1, the RAI wrestled with the question of whether an annual meeting created the kind of environment in which learned debate and knowledge exchange could be promoted. At its very first congress at Canterbury in 1844, Albert Way speculated that such an annual gathering was too large and too public a forum for the conduct of serious archaeological pursuits.[13] Few but their leading members had developed archaeological interests through formal academic channels and the intellectual credibility of its activities was far from a foregone conclusion. The collecting and assembling of artefacts for display at temporary museums at the summer meetings was thought to give the proceedings some scientific validity,[14] but the sense in the Institute was of continued concern over whether the meetings were really striking a meaningful educational note.

The focus of these concerns, it seemed, centred on the urban middle-class pleasure seeker for whom the meeting was more a recreational than a learning experience. Archaeological gatherings had every potential to draw in audiences of socialites, as the diary of Henry Crabb Robinson illustrates.[15] He recorded his attendance at the public dinner held on the occasion of the RAI meeting in Lincoln in 1848, writing that, 'I confess to all I meet, I make these journeys on account of the social pleasure I receive; and I perceive that it is because I give as well as take in this respect, that I am well received, though certainly one of the least learned Archaeologians who attend these meetings'.[16] Two other delegates who wrote a reminiscence of their attendance at the RAI congress at Bury St Edmunds in 1869 described their going as 'a pleasure trip, for it is on pleasure that we are bent'.[17] The Royal Archaeological Institute was

---

12 Society of Antiquaries Library, WAY SAL/MS/700/VIII, Ripon 1874.

13 Wetherall, 'From Canterbury to Winchester', p. 12.

14 Linda Ebbatson, 'Context and discourse: RAI membership, 1845–1942', in Vyner, *Building on the Past*, p. 31.

15 Crabb Robinson was an early nineteenth-century journalist and traveller. See Vincent Newey, 'Robinson, Henry Crabb (1775–1867), *Oxford Dictionary of National Biography* (Oxford, 2007), accessed 24 March 2012.

16 *Diary etc. of Henry Crabb Robinson* (1872 edition), quoted in Francis Hill, *Victorian Lincoln* (Cambridge, 1974), p. 80.

17 *Berrow's Worcester Journal*, 28 August 1869.

still preoccupied with the problem a decade later when a sub-committee, formed to discuss the efficacy of annual meetings, concluded that they 'were attended by too many people of whom too few were really antiquaries. They complained of the numbers of local tradesmen, newspaper correspondents and children, of the rush for carriages and meals, of the want of obedience to the director of excursions and of a lack of decorum when visiting private houses'.[18] It was a problem experienced at British Association events too. At Norwich in 1868, BAAS delegates attending a lecture by Dr Odling of the Royal Institution on 'Reverse Chemical Action', despite being treated to 'a number of brilliant and highly successful experiments', defied the formal courtesies of the meeting and made for the exit in large numbers before the talk was concluded so as not to be late for the following soirée. One visiting newspaper correspondent noted that 'the observations of the Mayor of Norwich in proposing a vote of thanks to Dr Odling were completely drowned by the bustle of a hasty departure', and concluded that the 'fashionable element' in the audience 'appeared more potent than the scientific'.[19]

Host towns and their organising committees were in the front line when it came to generating an audience and controlling access to the lecture halls, sectional meeting rooms and exhibition spaces of the annual meetings. Their task was a difficult one because, as we saw in Chapter 1, annual meetings served two different, and not necessarily complementary, functions. In one sense they were anniversary gatherings for existing members of the associations to reunite and renew their communications with the great and the good in the field. At the same time they were also recruitment events through which the associations hoped to gain new members, principally from the host locality. Success in attracting large numbers of learned visitors and incoming dignitaries from outside the immediate vicinity was often seen as a measure of the intellectual weight of a meeting. Special measures were sometimes put in place to achieve this. At Lewes on the English south coast, where the RASE show was held in 1852, five steam packets from France were promised, to provide agriculturalists from the Continent with the opportunity to attend.[20] Organising committees in the towns tried to find ways to demonstrate that they had succeeded in drawing in participants from a wide geographical area. A handsomely bound autograph book kept at Bristol for the signatures of delegates attending the 1836 British Association meeting there, provided space for each signatory to record their place of usual residence. The entries revealed 32 visitors from Wales, 11 from Ireland and dozens from throughout the south west, as well as a few from overseas, including attendees from Sweden, Russia and Tunisia.[21] Elsewhere, newspapers printed lists of learned visitors arriving in town to attend a meeting. In Glasgow during the NAPSS

---

18 Joan Evans, 'The Royal Archaeological Institute: a retrospect', *Archaeological Journal*, 106 (1951 for 1949): 1–11.

19 *Leeds Mercury*, 27 August 1868.

20 MERL, SR RASE/B/I/4, Minutes of Council, 1848–53 (rough copy), 7 May 1851.

21 BRO, 37454 (15)d, Autograph Book. British Association Autographs, 1836.

congress in 1860, the roll-call of distinguished names trumpeted by the editors of the *Glasgow Herald* included Judge Manockjee Cursetjee of Bombay, M.Y. Nagaoka of Japan, Mr Peabody, New York, the Rev. Dr Thomason, Berlin, and Mr W.M. Burke of Dublin.[22]

Where lists of distinguished foreigners could be complemented by high levels of local interest, a meeting could be said to have achieved both of its key purposes. At Newcastle in 1838 the BAAS congratulated itself both on the stimulus the meeting had given to local membership, but also on its capacity to draw in distinguished scientific visitors: '1246 new members have been elected at Newcastle ... The strangers coming are numerous and among them the best of the English, Irish and Scottish men of science ...'[23] A healthy recruitment of local attendees suggested that the evangelising mission of the association to stimulate learning in the locality had been achieved. The main instrument at the disposal of organisers to influence the character and number of people taking out membership locally was the ticket price. Different levels of payment secured different grades of access, as a notice printed in the several of the Manchester newspapers prior to the NAPSS meeting in the town in 1866 explained:

> Any person, lady or gentleman, becomes a member of the association by subscribing one guinea annually or ten guineas as a life payment. Members paying one guinea annually are entitled to attend the annual meeting of the association and to receive a copy of its transactions. Members paying two guineas annual or twenty guineas as a life payment have the additional privileges of attending the meetings of the association in London and receiving the publications connected therewith, with the use of the library and reading room. Any person, lady or gentleman, who pays ten shillings to the funds of the association is an associate for the annual meeting for which such payment is made, but is not entitled to the transactions.[24]

It was to this latter category of temporary members that many local people were admitted for the duration of an annual meeting. At Liverpool in advance of the BAAS' 1837 meeting, the local organising committee reported that 'Nearly 800 members have been admitted by the local council, six sevenths of whom probably are resident in Liverpool and the immediate neighbourhood'.[25] Such large increases in local membership at the time of an annual meeting were not universally welcomed, however. Strict admission criteria were imposed by the organising committee at Dublin before local people could be granted membership in advance of the BAAS

---

22 *Glasgow Herald*, 1 October 1874.

23 John Phillips to William Harcourt (10 August 1838), quoted in Morrell and Thackray (eds), *Gentlemen of Science. Early Correspondence*, pp. 272–3.

24 MCL, BR F306 S3, Social Science Association, Manchester Meeting, 1866. Notice printed in the *Courier*, *Guardian* and *Examiner*, 20 September 1866.

25 Bodleian Library, Dep. BAAS 17, Report of the Liverpool organising committee, 8 September 1837.

meeting in 1835. Recommendation by two existing members plus a monetary payment was required to protect 'the meeting from being totally overwhelmed by vast numbers of persons merely curious to witness an assembly which has far higher objects than to exhibit a pageant of science'.[26] These measures were put in place in response to advice from previous host towns where 'The admission was by much [*sic.*] too easy and the impression grows daily stronger that a money payment ... should always be demanded. It would appear probable that a £5 admission fee to be required of all new members would be a very effectual check upon ... the too easy admission of useless temporary members'.[27]

Reports of the most populous scientific events provide some insights into the concerns of organisers, at local levels and in the hierarchy of the associations, about the implications of liberal admissions policies. At Royal Agricultural Society shows, where mass attendances were the norm, one-shilling admission to the showground on designated days brought in a wide range of people whose keenness to interact with the livestock and exhibits sometimes caused disapproving comment and alarm. When the show was held in Battersea in 1862, nearly 40,000 people thronged the showground on the first of the three one-shilling days. At the livestock enclosure, crowds flocked to see the Shetland ponies and not all were content simply to look at the animals on display. One newspaper report described how 'The unfortunate little Shetlands huddled together – sires, dams and foals – were pinched, tickled, and not infrequently struck by the crowd assembled around them, and it was but seldom that the gaunt Scotchman in charge took the trouble to exert his voice in their defence'.[28] While levels of over-exuberance at the animal enclosures sometimes went unchecked, fewer chances were taken by local organising committees for RASE events over issues of public order and the safety and security of visitors. At Derby in 1843, for example, extra effort was made to guard against petty theft during RASE show week:

> There were present two officers from Leicester, two from Manchester, four from Sheffield, six from Nottingham, and twenty (of the A division) from London, besides our own police force. We heard of no serious offence being committed by the thieves, who are sure to congregate in larger or smaller numbers on such occasions. Some few, however, were apprehended for attempting to commit robberies on the person, one of them, being a man named George Gorthorn, a well-known character in Nottingham, Leicester and Derby.[29]

It was not just to deter the activities of habitual thieves that a police presence was sometimes required during annual meeting weeks. The prospect of glimpsing a

---

26 Bodleian Library, Dep. BAAS 17, Report of the local council of the Dublin meeting, 8 August 1835.

27 Bodleian Library, Dep. BAAS 133, Notebook of John Phillips, 1835.

28 *The Era*, 6 July 1862.

29 *Derby Mercury*, 19 July 1843.

celebrity visitor was enough to prompt crowds of otherwise 'respectable' inhabitants to assemble in defiance of the authorities. At Brighton in 1872, the arrival of the Emperor Napoleon at the town's Grand Hotel during British Association week sparked considerable excitement.[30] According to one report, 'several hundreds of highly respectable ladies and gentlemen', many of them equipped with opera glasses, took up positions on the steps of the hotel hoping to catch sight of the Emperor. When they refused to disperse, the police were called and, amid much bitter complaining, the onlookers were moved on.[31] Where large numbers were involved, crowd control sometimes proved a challenge for those charged with the task of officiating at meetings. At Bath in 1864, for example, the inclusion in the programme for BAAS meeting week of a lecture by the African explorer David Livingstone generated great interest. The 'celebrity' explorers of the mid-nineteenth century had great crowd-pulling potential and often made Section E, Geography, the most populated division of the British Association meetings.[32] In an attempt to meet the demand to hear Livingstone's lecture at the town theatre, organisers hastily arranged for a simultaneous reading of the paper at a different venue so that more people could be accommodated. Despite these provisions, it was still necessary to put in place robust crowd control measures at the theatre where 'barricades of a substantial character were erected around the building, and in some places they were double ... When the doors were thrown open, a general rush ensued. Pit boxes, gallery, orchestra and stage were crowded in a trice and never has there been such pushing and squeezing as to witness the debut of the African 'Lion' on the Bath boards'.[33]

Livingstone was not the only scientific luminary whose presence at an annual meeting could produce chaotic scenes in a lecture theatre or meeting room. In the late 1860s when Thomas Henry Huxley was at the height of his fame and renowned for voicing his views on evolution and its social and moral implications, his appearance at British Association meetings caused great excitement.[34] At Exeter in 1869, people crowded into sectional meeting rooms to hear him give even the briefest contribution to a discussion. The *Bradford Observer*'s correspondent wrote that, 'Yesterday, a paper was read in the Zoological Section advocating the legislative enactment of a "close time" for birds. Professor Huxley rose to

[30] Louis-Napoleon Bonaparte (1808–1873).

[31] *Daily News*, 15 August 1872.

[32] Charles W.J. Withers, *Geography and Science in Britain, 1831–1939. A Study of the British Association for the Advancement of Science* (Manchester, 2010), p. 92. Withers also notes that the precise designation of Section E altered during the second half of the nineteenth century as Geography was combined with Geology and later Ethnography, p. 68.

[33] *Bristol Mercury*, 24 September 1864.

[34] For a recent account of his debate with Wilberforce on Darwinian Theory, see Frank A.J.L. James, 'An "Open Clash between Science and the Church"? Wilberforce, Huxley and Hooker on Darwin at the British Association, Oxford, 1860', in David M. Knight and Matthew D. Eddy (eds), *Science and Beliefs. From Natural Philosophy to Natural Science, 1700–1900* (Ashgate, 2005), pp. 171–91.

make some remarks on the subject, and the meeting rapidly filled. During a five minutes' discussion at least a hundred people entered the room'.[35] Such audience behaviour can hardly have been what organisers had in mind when they drew up arrangements for the sectional meetings. (The same correspondent pointed out that the different sections were not closely located for convenient movement between them.) Nevertheless, it demonstrated that even in the formal scientific surroundings of the sectional meeting room people were sometimes drawn as much by a desire to indulge their fascination with the well-known personalities of the meeting as to engage with the ideas and theories advanced by speakers.

If the conduct of fee-paying visitors to show grounds and meeting rooms sometimes caused concern and defied and frustrated the expectations of speakers and organisers, so too did standards of delivery and organisational arrangements occasionally fall short of delegate expectations, leading to frustration, fatigue and boredom among audiences. At the Archaeological Institute's meeting in Salisbury in July 1849, the audience in the Architectural Section was short-changed when Mr John Britton stood up to announce that he had prepared a paper on the history and origin of market crosses, with particular studies of those in Glastonbury, Salisbury and Malmesbury, only to confess that 'in the absence of the drawings necessary to illustrate the paper (which were accidentally forgotten), he must postpone the reading of the paper to a future evening'.[36] A combination of poor delivery and extraneous noise spoiled Sir William Beetham's paper on the origin of idolatry at the Canterbury meeting of the archaeologists in 1844. His talk was 'but very imperfectly heard, owing to the low tone of voice in which it was hastily read, and to the frequent interruptions from voices in the street. It was received with applause, not unmingled with occasional laughter'.[37] At Bristol in 1898 the BAAS meeting was held during a week of unusually warm weather which did nothing for the comfort and attention spans of the delegates who packed into crowded meeting rooms: one visitor complained that 'The president's address was learned, long and dreary – his figures were appalling', and that 'In tropical heat we endured the Biological meeting at the zoo'. The British Association, he observed, was, 'yearning to shower draughts of selected science on the perspiring population, the majority of whom are weakly thirsting for copious draughts of another sort in the paralysing heat of this glorious Summer'.[38] The tendency of presidential addresses to test the concentration levels of even the most devoted of attendees was a problem which afflicted the other associations too. George Hastings wrote to Lord Brougham in September 1860 in advance of his arrival at the Social Science meeting to be held that year in Glasgow, and advised him to edit the content of his presidential address as 'two hours is too long to keep

35 *Bradford Observer*, 23 August 1869.

36 *Hampshire Advertiser and Salisbury Guardian*, 28 July 1849.

37 SAL/MS/877/1, British Archaeological Association, cutting of report by *The Times*' correspondent in Canterbury, n.d.

38 BRO, 29427/3, '1898 Bristol British Association. Notes by a Cheerful Pterodactyl'.

an audience in the evening and with speeches after'.[39] For ordinary delegates there was sometimes a sense of frustration at feeling remote from the dignitaries and scientific notables assembled at the meetings. One participant at the Belfast meeting of the BAAS in 1852 wrote to the Earl of Rosse to express his regret at not having seen him there, recounting an anecdote 'of an old woman who went to hear Whitfield when he was in the height of his celebrity and attracted vast crowds. She could not get near enough to hear a word he said; but expressed her high satisfaction at having "caught a glimpse of his godly wig"! Such was a great part of what I was forced to content myself with!'[40]

If the delegates inside the meeting rooms and lecture theatres occasionally felt remote from the proceedings, it is tempting to conclude that the wider urban populace must have been entirely disengaged. But here, the role of the press in reporting annual meetings was central to the process of engaging the wider, non-attending public. Not only did the provincial press take a keen interest in annual meetings as urban events, they also covered the proceedings in the sectional meetings and the general lectures and addresses at congresses of the BAAS and other learned associations in great detail. As the editorial in the *Newcastle Chronicle* put it, 'During the ten days the association has been in session we have published double numbers of the *Chronicle* and, with the view of giving the fullest publicity to the reports of its proceedings we have during that time, to a great extent suspended the news of the world. For the last fortnight we may be said to have fed the population of the district upon pure science'.[41] Where lectures were deemed to be especially popular and of interest to a wide general audience, coverage was especially full. Such was the case at Oxford in 1860 when Captain Parker Snow read a paper to Section E, Geography and Ethnology, on the lost polar expedition. *Jackson's Oxford Journal* judged this to be the most popular section because 'the papers read there have been of general interest, and such as have been intelligible to a mixed audience', and also because 'many of the distinguished persons who have introduced subjects, or spoken upon them, are men whose names and daring deeds have become as a familiar as household words'. Consequently, the editor decided to devote as much space as possible to covering the proceedings of the section, 'in order that the public generally might participate in the pleasure and interest which have characterised it day by day'.[42]

In addition to the profusion of newsprint dedicated to the proceedings, working inhabitants of the host towns also had opportunities to access the sectional rooms and lecture halls in person during annual meetings, but these were by

---

39 UCL Special Collections, B MSS, 7663, George W. Hastings to Lord Brougham, 18 September 1860.

40 Bodleian Library, Dep. BAAS 142, R.D. Dublin to the Earl of Rosse, 16 September 1852.

41 Bodleian Library, Dep. BAAS 147, Printed material for the annual meetings, 1860–63. Report of meeting at Newcastle in 1863 from *The Chronicle* (undated).

42 *Jackson's Oxford Journal*, 7 July 1860.

no means equal across the four associations. The Social Science Association went to considerable lengths to attempt to engage working-class townspeople with the business of their meetings and used formal channels of communication with provincial mechanics institutes, for example, to achieve this. NAPSS rules allowed mechanics' institutes and working men's colleges to affiliate themselves for a subscription of one guinea per year, and send two representatives to annual meetings, but the response to their invitation was far from overwhelming. Hastings informed Lord Brougham in 1858 that 'A letter was written by me early in the spring to every mechanics' institute; I obtained very few replies, and not more than half a dozen (at the most) subscribed. It is earnestly to be desired that this Rule of the Association should be availed of – and thus working men be brought to the meetings'.[43] The relative ineffectiveness of these channels in generating a response may simply have reflected the difficulties faced by many mechanics' institutes by the second half of the nineteenth century when falling membership levels were forcing many to channel their efforts into the promotion of broader recreational, rather than purely educational, activities.[44] It did not, however, signal a lack of interest on the part of working people in the debates of the day. When the temperance campaigner and publisher John Cassell[45] inaugurated a prize essay competition for operatives, their wives, daughters or sisters to express their views on social questions, over 500 entries were submitted, many on the subject of 'Self-Education'.[46] The prizes, which were awarded at a soirée in the Bradford Mechanics Institute during the Social Science Association's meeting in that town in 1859, went to a range of recipients including a carpenter, gun engraver, baker, factory operative, plumber, gardener, dyer, shoemaker, clothiers' cutter, carpet weaver, china painter, shipwright, compositor and to a number of women including the wife of a blacksmith and the daughter of a working man.[47]

Arguably, the NAPSS was more successful than the other associations under scrutiny here in accommodating working-class delegates and catering to their needs. Meetings specifically for working men were instituted at some of the earliest annual congresses of the association at Liverpool and Bradford, and it became an important measure of success for the annual meetings of the NAPSS that working people had not only had the opportunity to attend but had done so with enthusiasm and in large numbers. In his concluding remarks at the end of the Manchester meeting in 1866,

---

43 UCL Special Collections, B MSS 13, 151 George W. Hastings to Lord Brougham, 1858.

44 Gerry Wright, 'Discussions of the characteristics of mechanics' institutes in the second half of the nineteenth century: the Bradford example', *Journal of Educational Administration and History*, 33/1 (2001): 2.

45 Rosemary Mitchell, 'Cassell, John (1817–1865)', *Oxford Dictionary of National Biography* (Oxford, 2004), accessed 29 February 2012.

46 See *Social Science: being selections from John Cassell's prize essays by working men and women with notes* (London, 1859), p. ix.

47 *Newcastle Courant*, 28 October 1859.

Hastings particularly noted the success of the Friday night meeting of working men, although his description of the audience 'beaming with intelligence and good humour, kindness and respect, coming to be instructed with a full intention of turning what they learned to the advancement of their position, to the benefit of their families and to the good of the country'[48] may have been designed more to flatter the Manchester organisers than to reflect accurately the conduct of the attendees. More boisterous levels of participation were also sometimes in evidence, as at Glasgow in 1860 when Lord Brougham's appearance at the City Hall for the working men's meeting 'was the signal for general cheering, which lasted several minutes, the company rising *en masse* and saluting his lordship by waving of hats and shouting loud hurrahs'.[49] It was also possible at meetings of the Social Science Association for the working classes to participate in debate in the sections. Hastings noted that in Liverpool in 1858, 'a number of working men took part in the discussions. I do not mean at the working men's meeting but in the departments. Several read papers – chiefly on cooperative societies, which appear to be a remarkable feature in the social history of the present day'.[50]

In contrast, the British Association was less adept at attracting working-class delegates to its meetings, even where the subjects of discussion directly concerned them. Evening meetings were laid on so that ordinary townspeople could access some of the scientific business after the end of their working day. For Adam Sedgwick, who delivered an evening lecture at the Manchester BAAS meeting in 1842, it was a worthy endeavour but one which did not engage the majority of the meeting's delegates. He said of the working men he met that evening:

> I found upon them the marks of intellectual minds, and the proofs of high character; and I conversed with men who, in their own way, and in many ways bearing upon the purposes of life, were far my superiors. I would wish the members of the British Association to mingle themselves with these artisans, in these perhaps overlooked corners of our great cities.[51]

In Bradford in 1873, the BAAS' Economic Science and Statistics Section debated the cost of living and 'the effect of the increase of prices on certain necessaries of life'. In the course of the discussion Lord Houghton, 'after expressing his opinion that the subject was one more suited to the Social Science Congress than to the British Association, expressed his regret that they had not present a fuller representation of the working classes'.[52] In reality, the high cost of attending – the

48 MCL, BR F306 S3, Social Science Association, Manchester Meeting 1866, cutting from *Manchester Guardian*, October 1866.

49 *Daily News*, 28 September 1860.

50 UCL Special Collections, Brougham MSS, 7656, George W. Hastings to Lord Brougham, 21 September 1859.

51 Clark, *The Life and Letters of the Rev. Adam Sedgwick*, vol. II, p. 46.

52 *Birmingham Daily Post*, 20 September 1873.

cheapest available admission was £1 to become an associate member for the year with access to sessions at the current annual meeting[53] – put the event beyond the reach of a working-class audience. Also problematic was the simple fact that much of the business of Social Science congresses and British Association meetings took place during the working day when most ordinary people were simply not free to attend. In early August 1848 a local resident wrote to the editors of Swansea's *Cambrian* newspaper lamenting the fact that local apprentices and assistants were effectively excluded from attending the forthcoming meeting of the British Association in the town because the bulk of meetings and lectures took place during normal business hours.[54] Similar sentiments were expressed by 'one interested', an inhabitant of Manchester, who wrote to the *Manchester Guardian* in 1866 on the eve of the Social Science congress there to complain that:

> I, in common with the majority of my fellow citizens, am very actively engaged in business during the day time ... Consequently we have no opportunity of attending the majority of the meetings of the various sections. There are, however, the evenings, but here we are debarred by the arrangements of the committee for unless a ticket is taken for the whole course there is no admission to an evening meeting.[55]

Where evening events were open to all comers, take-up was often high. At Aberdeen in 1859, an exhibition of portraits and autographs illustrating the civil and literary history of Scotland, as well as of Aberdeen itself, proved very popular with the inhabitants of the district. 'It is worthy of remark' wrote the *Aberdeen Journal*, that 'the whole suite of rooms containing the objects was often crowded at night by the working classes, many of them accompanied by their children', and it added approvingly that 'not a single article was injured or displaced'.[56]

Not all elements of the annual meeting were confined to halls and meeting rooms and demanded paid admission. With local organisers incorporating civic ceremonial and festive street decoration into the proceedings, it was possible for local people simply to turn out in host towns to see the sights. In Manchester in 1869, when the Royal Agricultural show was in town, with the Prince of Wales in attendance, ordinary townspeople embraced the festivities enthusiastically: 'Some of the omnibuses were studded with flags, tied with "red, white, and blue" rosettes; and even the needy knife-grinder, as he pursued his humble trade in the suburbs, stuck a small tricolour over his hand-machine'.[57] At Bury St Edmunds in 1867, the street lights and ornaments put up for the RASE show provided a point of interest for local people and visiting show-goers alike. Local reports observed how, 'As evening closed in the townspeople generally joined the visitors in the streets, in

---

53 *Leeds Mercury*, 17 September 1873.

54 *Cambrian*, 4 August 1848.

55 MCL, BR F306 S3, *Manchester Guardian*, 4 October 1866.

56 *Aberdeen Journal*, 6 June 1860.

57 *Daily News*, 21 July 1869.

order to see the illuminations and decorations'.[58] At Dundee in 1867, news of an electric light displayed in front of the High School where several BAAS sections were housed during the British Association's annual meeting that year brought out crowds of onlookers. Despite their curiosity, these sight-seeing locals were not easily impressed with the faint, flickering spectacle, and it was reported that 'Large crowds witnessed the display, but almost all seemed to have formed the idea that the light was to be a steady, continuous one, and illumine the whole street, and seemed much disappointed at the result'.[59]

The crowds that gathered on such occasions were rarely made up of town dwellers only. Prominent among their number were visitors from the surrounding countryside. Towns were a magnet for rural visitors at holiday times, especially at Whitsun, or annual events, such as race meetings or fairs, and scientific congresses, too, brought the potential for rural-urban encounters. The *Devon Weekly Times*' journalist, in Exeter for the British Association meeting in 1869, described train loads of 'our country cousins' from surrounding districts, flocking into the town for the occasion. Many of them, he claimed, 'bent their steps to the Cavalry Barracks, to see the flowers and the dogs', or stood, like Farmer Diggins of Dartmoor, 'looking about the city in amazement' and enquiring how he might recognise the 'wise men' among the visitors on the streets.[60] This image of the innocent, pleasure-seeking 'country cousin' was common in the provincial newspapers of the day which peddled stereotypes of cheerful farming people with money to spend, drawn to the attractions, but also vulnerable to the vices of the town.[61] It offers little real sense of how the populace of the British countryside responded when one of the major knowledge congresses of the day was held in their vicinity, especially those who were fee-paying attendees. At the RASE show in Bury St Edmunds in 1867, country visitors were credited with knowledgeable engagement rather than simply pleasure-seeking intent, especially the agricultural labourers who were 'quick in their discovery and keen in their appreciation of the most striking novelties' of the assembled machines for threshing, hay-making, chaff-cutting and root-pulping.[62]

For those without direct access to the enclosed spaces of meeting room or showground, the visiting scientists were themselves a spectacle to be viewed and admired, even by those who had little sense of why they were gathering. When the archaeologists descended on Canterbury for their annual meeting in 1844, the correspondent for *The Times* newspaper noted that the antiquaries' arrival had created great excitement, although there was confusion as to whether the influx of visitors into the town was due to an early sitting of the assizes or a race meeting. Nevertheless, he observed that, 'from the oldest inhabitant of the city to

58 *The Bury and Norwich Post and Suffolk Herald*, 23 July 1867.

59 *Dundee Courier and Argus*, 6 September 1867.

60 *Devon Weekly Times*, 20 August 1869.

61 See for example, *Derby Mercury*, 3 July 1850; *Penny Illustrated Paper and Illustrated Times*, 3 August 1872; *Aberdeen Weekly Journal*, 5 October 1878.

62 *Bury and Norwich Post and Suffolk Herald*, 23 July 1867.

the youngster at the National School, through the whole three wards, inside and outside the walls, all is unusual wonderment'.[63] At Swansea in 1848, where BAAS visitors arrived by sea, crowds congregated at the quayside to view their arrival:

> At the Mumbles point, where the passengers were landed, there were crowds assembled to see the philosophers pass, and along the whole road to Swansea there were groups here and there collected to gaze. The street of Swansea itself, near the Royal Institution … was lined with people, and policemen were in attendance to clear the way.[64]

In Nottingham in 1866, local youths exhibited similar levels of interest in the movements of BAAS delegates about the town. The women's suffrage campaigner, Kate Stanley, wife of John Russell, Viscount Amberley,[65] attended the meeting with her husband and recorded in her journal the attention she and her party attracted as they made their way through the town between meeting venues. She recalled that, 'As we walked through the streets, we were followed by a crowd, chiefly of dirty boys, and now and then cheered. M. de Tchihatchef who was with us took it at first to himself as being a foreigner and bowed to the mob'.[66]

Chance encounters such as this between townspeople and attendees, though rarely documented, must have been a feature of most meeting weeks. The fact that Viscountess Amberley recorded the incident in her diary, between accounts of papers and discussions in the sectional meetings, suggests that it had an influential effect on her experience of the event. But other fragments of evidence reveal that it was not just the visiting scientists who formed an object of fascination for locals: the urge to encounter and observe was reciprocal. William West of Leeds, an attendee at the BAAS meeting in Cork in 1843, approached the congress in a Mayhew-like spirit of social investigation and wrote to the editors of the *Leeds Mercury* about his encounters with and impressions of the local population. He was struck by the irrepressible humour of the people, like the female apple-seller who called out to him and his party as they made their way to their hotel: 'how did you leave all your friends on the other side of the water, my darling?'[67] But he also sought out opportunities to interact with the lower classes of the locality, once going into a 'hovel' occupied by a women and her six 'barefooted children' and

---

63 SAL/MS/877/1, British Archaeological Association, cutting from report by *The Times*' correspondent in Canterbury, n.d.

64 *Morning Post*, 11 August 1848.

65 Ann P. Robson, 'Russell, Katharine Louisa, Viscountess Amberley (1842–1874)', *Oxford Dictionary of National Biography* (Oxford, 2004), accessed 29 February 2012.

66 Kate Stanley's Journal, Thursday 23 August 1866, in Bertrand and Patricia Russell (eds), *The Amberley Papers. Bertrand Russell's Family Background*, vol. 1 (London, 1937), pp. 524–6. Pierre de Tchihatchef was a Russian-born geologist and naturalist who addressed a number of British Association meetings in the 1860s.

67 *Leeds Mercury*, 26 August 1843.

interviewing her about her husband's employment and earnings. He hired local 'gossoons'[68] to perform menial tasks for him during his visit. The idea of this was 'more … to study them than to relieve me of the light labour of carrying my telescope and note-book'. On the basis of these contacts, and probably equally influenced by the negative stereotypes of the Irish poor which abounded in Britain in the 1840s,[69] he concluded that they were a people characterised by 'gratuitous mendacity' and 'general slovenliness and wretchedness'.[70]

West's initiatives to put himself in the path of the most impoverished sections of the local population may not have been typical of the behaviour of delegates at these meetings, but they demonstrate the capacity of the host town and its population to form part of the subject matter ripe for study by eager scientific visitors. The people encountered by West, Stanley and other attendees as they acquainted themselves with their surroundings during meeting week did not constitute a scientific audience in any conventional sense. Evidence of their interactions with the learned visitors very rarely finds its way into the written

Figure 4.1 Two Newhaven fishwives encountered by a visitor to Edinburgh for the Royal Archaeological Institute Meeting in 1856, *Illustrated Times*, 9 August 1856 (author's collection)

[68] A gossoon was a servant-boy or lackey. See 'gossoon, n.', *Oxford English Dictionary* (online edition, June 2011), accessed 27 September, 2011.

[69] See for example, Michael W. De Nie, *The Eternal Paddy: Irish Identity and the British Press, 1798–1882* (Madison, 2004).

[70] *Leeds Mercury*, 26 August 1843.

record of the congresses, but the responses of these people to the sights and sounds of the meeting and to the presence of visitors in their midst indicates the breadth of impact which these events had on the localities, and provides an important reminder of the range of people who experienced an annual meeting in some form.

Local organising committees were acutely aware of the likelihood of such encounters and in much pre-meeting rhetoric there was a palpable sense of them trying to instil a sense of shared responsibility for the smooth running of the meeting into the community as a whole. Even members of the press, previously critical of the worth of the learned associations, pleaded the case for co-operation by all concerned for the overall good of the town. The *Manchester Examiner* on the eve of the first NAPSS meeting in that town in 1866 was a case in point:

> Hitherto we have contemplated the movements of this body from a distance which, if not always perfectly respectful, had at least the advantage of permitting us to be irreverent with impunity. Now the case is altered. The lights of social science are coming here. The sciences and humanities will be before us in bodily shape, and the moral terrors they are calculated to inspire will compel us to change our manners. The members of the Association will be amongst us as visitors and invited guests. Our citizens for the first time will have to perform the duties of hosts and we who speak in their name must at least be obliging and civil.[71]

It must have been with a sense of relief as well as local pride that George Hastings' concluding remarks at the end of the meeting were heard to include thanks to Manchester's industrial population for their 'demeanour, not only at meetings but also in the streets'.[72]

It was not only the behaviour and level of engagement of working-class people and the general populace in annual meetings which was keenly discussed and debated. Women were another section of the population whose participation at the congresses was a subject of frequent comment. All four of the associations under scrutiny here admitted women to their annual meetings, but the levels of involvement which the females in attendance were expected, or encouraged, to aspire to varied greatly. The most comprehensive study of women at BAAS meetings has noted that they primarily consisted of relatives of the visiting scientists, members of the host town's leisured classes or tourists, and that their admittance at least to dinners and conversaziones was 'never questioned'.[73] In fact, this was not always the case. Ladies' admission to the presidential dinner at the BAAS meeting in Cheltenham in 1856 was described as 'a novel feature … the presence of the fair sex at public dinners being hitherto

---

71 MCL, BR F306 S3, Social Science Association, Manchester Meeting 1866, cutting from *Manchester Examiner*, 20 September 1866.

72 MCL, BR F306 S3, Social Science Association, Manchester Meeting, 1866, cutting from the *Manchester Guardian*, October 1866.

73 Higgitt and Withers, 'Science and Sociability', pp. 6–8.

considered not quite *comme il faut*'.[74] With traditions of all-male dining in other institutions, notably the Royal Academy,[75] it is perhaps not surprising that eyebrows were raised in some quarters at their admission to BAAS banquets. In general though, women's presence at the social elements of annual meetings was encouraged as a kind of embellishment which might visually and socially enhance the proceedings. At Aberdeen in 1859 it was anticipated that ladies would 'find the evening meetings and conversaziones congenial, and will, in turn, impart grace and attraction to them'.[76] At Bury St Edmunds when visitors gathered for the annual meeting of the Archaeological Institute in 1869 the Abbey grounds provided an evening venue. It was noted by one local newspaper correspondent that, 'There was a good sprinkling of ladies to relieve the "Dry as Dusts" and certainly the Abbey precincts rarely looked so attractive as on this occasion'.[77] This 'ornamental' female presence was all well and good so long as they did not become a distraction from the serious scientific business of the meetings, and organisers sought to keep tight control of all aspects of female participation. At Bristol in 1836 the local committee charged with making arrangements for the British Association visit seemed to be attempting to limit the visual spectacle that a fashionable female presence might create at the meeting when they stipulated that 'ladies will be expected to attend the evening meetings in plain evening dresses without bonnets'.[78] Their principal means of controlling female involvement, however, was through the allocation of ladies' tickets. Too free an admittance of women was seen by the BAAS as having a potentially diluting effect on meetings. During preparations for the Dublin congress in 1835, recent practices at Edinburgh and Cambridge were reviewed and 'The admission of ladies was supposed to have been too indiscriminate'.[79] In some quarters women were viewed as a financial burden on the meeting, especially where catering and hospitality costs were concerned. They were accused, on some occasions, of over-indulging in the refreshments on offer at evening gatherings. The *Bradford Observer*'s correspondent at the BAAS meeting at Exeter in 1869 lamented the behaviour of ladies at the refreshment table at the conversazione where, 'Last night a number of them occupied the counter for long periods, sipping repeated glasses of iced claret and exhibiting a healthy appetite for cakes and sweets'. He claimed that 'refreshment contractors will cater for a party of gentlemen only at a lower charge per head than for a mixed party of ladies and gentlemen'.[80]

---

74 *Morning Chronicle*, 30 July 1856. *Comme il faut*, adj. and adv., 'as it should be', *Oxford English Dictionary* (online edition, June 2011), accessed 29 February 2012.

75 Holger Hoock, 'From Beefsteak to Turtle: Artists' Dinner Culture in Eighteenth-Century London', *Huntington Library Quarterly*, 66/1–2 (2003): 38.

76 *Aberdeen Journal*, 31 August 1859.

77 *Ipswich Journal*, 24 July 1869.

78 BRO, 32079 (39) Minute Book of the Reception Committee for the British Association visit, 15 August 1836.

79 Bodleian Library, Dep. BAAS 133, Notebook of John Phillips, 1835.

80 *Bradford Observer*, 23 August 1869.

Figure 4.2 Refreshments at a British Association soirée in the town hall at Birmingham in 1865, *Illustrated London News*, 23 September 1865 (author's collection)

This criticism of women's behaviour in social gatherings at scientific meetings, however, was mild in comparison to the censure sometimes directed towards those who participated in scientific elements of the congresses. For a period in the middle of the nineteenth century, press reports containing withering accounts of uninterested, ill-informed women taking up valuable space in meeting rooms were not uncommon during meeting weeks. In the British Association in particular there was an evident belief among a portion of the male scientific fraternity that space in the sectional meetings was wasted on provincial urban women. One member of the Newcastle organising committee in 1838 claimed that the women who attended the annual meeting at Bristol the previous year were 'the wives and daughters of wealthy merchants and tradesmen, who had never looked even into the title page of a book of science, who could not tell you the difference between *geo*logy and *genea*logy; or *ast*ronomy and *gast*ronomy'. He was no more complimentary about the prospective women attendees of Newcastle: 'Are there 500 ladies living in Newcastle, who can enter with zest into the *scientific* portion of the week's proceedings? I boldly affirm that there are not'.[81] During the meeting itself, the *Morning Post*'s correspondent in Newcastle joined in the general denunciation of the females present, complaining about 'the everlasting shuffle, scrape, scratch, and shifting of the promenade of ladies on the bare boards – some tripping up

---

[81] *Newcastle Courant*, 20 July 1838.

stairs in reckless haste to indulge their curiosity, and disturbing row after row of real listeners'.[82] This attempt to distance women from the real scientific audience reflected more than just the prevailing gender ideology of the day. In an era when scientific disciplines like Botany, popular with women and men alike, were attempting to legitimise themselves as serious pursuits rather than polite pastimes, women did not fit the image of the expert practitioner and their presence became problematised. Their scientific activities were better confined to the reading matter of the domestic drawing room or the naturalists' society field excursion, leaving the laboratory and lecture hall to the (male) specialist.[83]

Although never excluded from the parliaments of science, the separate or special arrangements put in place for women who wished to attend can be understood in this context. Local organising committees often made particular provision for female delegates, setting aside separate seating areas in meeting rooms as at the BAAS' Liverpool meeting in 1837, when the general council was notified that 'the galleries of the Royal Institution and in the Mechanics Institution should be reserved for the accommodation of ladies at the sectional meetings'.[84] The RAI's visit to Newcastle-upon-Tyne in 1852 included an excursion to Hexham and the Roman Wall. The organisers clearly expected a number of women to be in attendance and undertook to 'provide such conveyances as he can for those ladies who cannot walk the whole distance'.[85] The excursions and site visits, conducted during the course of meeting weeks, often attracted large numbers of women. At the archaeologists' first annual meeting in Canterbury in 1844, members, among them a number of women, were given the opportunity to attend the site of an excavation. Reporters commented approvingly on the ability of the women in attendance to endure adverse weather conditions for the sake of extending their knowledge:

> Though the weather was very unpromising, from 150 to 200 members and ladies set out about ten o'clock for Breach Down, about eight miles off, where the first barrows were pitched upon for opening. Hardly had the operation commenced, when a deluge of rain descended; but such was the excitement of the scene, that many of the more delicate sex retained their situations and would not lose sight of the interesting excavations. It was indeed delightful to notice the feeling with which our fair countrywomen made for once participators in an intellectual pursuit with their husbands, fathers, brothers and friends, examined every ancient memorial disinterred from the universal mother earth.[86]

---

82 *Morning Post*, 5 September 1838.

83 See for example, Ann B. Shteir, *Cultivating Women, Cultivating Science: Flora's Daughters and Botany in England, 1760–1860* (Baltimore, 1996), pp. 147–69.

84 Bodleian Library, Dep. BAAS 17, Report of the Liverpool organising committee, 8 September 1837.

85 SAL, WAY MS/790, RAI Meeting at Newcastle 1852.

86 SAL/MS/877/1, 1844–47, *The Literary Gazette and Journal of the Belles Lettres, Arts, Sciences &c.*, London, 14 September 1844.

The critical *Times*' correspondent gave a rather different description of the same event and clearly thought the women rather out of place, claiming that although the antiquarians were oblivious to the rain, 'ladies ran to windmills and little thatched cottages; and the last new fashion from town, green, blue and red muslins, were seen mingling with the flour sacks of the miller and the home spun or Manchester cottons of the humble cottager'.[87]

With these sentiments colouring much of the press reporting of women's participation in annual meetings, diary evidence of women attendees takes on relatively greater significance and in some cases presents a very different picture of the female experience at these events.[88] By no means all women confined their participation to appearing at social functions. Engagement with the scientific business of the lectures, addresses and sectional meetings was occasionally communicated in the diary records and private correspondence of women attendees. Kate Stanley, who attended the British Association's annual meeting in Nottingham in 1866 kept a daily record of the meetings, lectures and social functions in which she participated with other members of her family. Her brief references to the papers she heard and the conversations she had with scientists outside the meeting rooms reveal her to have been a well-informed and engaged listener. On 23 August she heard a paper by Sir Samuel Baker on the Nile sources[89] which she described as 'very interesting' and commented that, 'though I had read his book there was a good deal of new stuff'.[90] She also attended sectional meetings separately from her husband, as on Friday 24 August when she chose the proceedings of Section E in the Mechanics' Hall while Viscount Amberley was in the Biology section.[91] Other female attendees who elaborated on the scientific papers at BAAS meetings included Rosetta Flower, wife of Sir William Flower, who wrote to her mother in 1862 with an account of her attendance at the Cambridge BAAS meeting that year. Although much of her description of the event dealt with the setting and the people in attendance, she gave a particularly animated account of her attendance at Section D (Zoology, Botany and Physiology) where the debate between Huxley and Owen was 'thronged to excess' by people who were 'attracted partly by Owen's name and partly by expectation of a battle'. For her part she found the exchanges 'most exciting and interesting', and took particular pride in hearing her husband

---

87 SAL/MS/877/1, 1844–47, cutting 'from our own correspondent', n.d.

88 Diary evidence is deployed extensively by Charles Withers and Rebekah Higgitt in their study of women at BAAS meetings. See 'Science and Sociability', pp. 1–27.

89 Sir Samuel Baker's Nile expedition, 1861–65, resulted in findings which linked the river's sediments to tributaries in Abyssinia. Thomas P. Ofcansky, 'Baker, Sir Samuel White (1821–1893)', *Oxford Dictionary of National Biography* (Oxford, 2004; online ed. 2008), accessed 3 March 2012.

90 Kate Stanley's Journal, Thursday 23 August 1866.

91 Ibid., Friday 24 August 1866.

respond to some of the points raised.[92] Like Rosetta Flower, William Pengelly's wife, writing to her sister from the British Association meeting at Plymouth in 1877, commented not just on the social functions and parties she attended there, but also on the scientific proceedings: '… many of the Papers are interesting. I enjoyed one to-day by Professor Rolleston exceedingly. The subject was, "The Flora and Fauna of Prehistoric Times". He also made some very good remarks on the various objects of interest in the Plymouth Museum, brought here for exhibition'.[93]

Figure 4.3 Female delegates at the 1879 Social Science congress in Manchester, *The Graphic*, 11 October 1879 (author's collection)

92 Dep. BAAS 142, Rosetta Flower (daughter of Admiral Smyth and wife of Sir William Flower) to her mother, written at Cambridge, 3 October 1862.

93 Pengelly (ed.) *A Memoir of William Pengelly of Torquay, FRS*, p. 251.

Despite the common presence of these women at the proceedings by the late 1860s and 1870s, some sections of the provincial press were slow to adapt. It was still a matter for surprised comment when, as at the BAAS' Norwich meeting in 1868, the women present at the presidential address 'appeared to manifest a greater amount of interest in the proceedings than they could have been expected to give the discussion of scientific subjects'.[94] For some journalists, the way to approach the involvement of women in the scientific business of the meetings was to portray them in less than flattering terms as unfeminine, unattractive, political radicals, or as *The Devon Weekly Times*' correspondent in Exeter for the BAAS meeting in 1869 put it, as 'learned ladies in spectacles – the champions of women's rights'.[95] But by this period, attitudes had shifted markedly within the associations, notably the BAAS. The botanist Lydia Becker, who presented five papers at British Association meetings between 1868 and 1874, was 'the most visible woman participant in its first half century'.[96] Her lectures attracted large audiences and addressed issues of gender difference and women's capacity for scientific learning. At the Bristol BAAS meeting in 1875, papers by Miss Buckland in the Anthropology Section and by Mrs Crawshay in the Economic Science and Statistics Section drew interested audiences, many of whom were themselves women. At the Anthropology Section in the Royal Hotel, one journalist indignantly described how, 'the reporters' table was swept away to afford room for the ladies, who charged like an impetuous squadron of light cavalry into the few square yards of space left by the retreating newspaper men'. Worse still, at the Economics and Statistics Section where Mrs Crawshay's address on 'Domestic Service for Gentlewomen'[97] was the fifth paper on the programme, 'impatience was openly manifested' by a large audience, 'ladies chiefly', who objected to the length of the preceding address on sugar duties. 'So there were cries of "time" and counter cries of "order"' until Mrs Crawshay finally took the platform to general applause.[98]

Part of the explanation for the shift in levels of involvement of women in BAAS meetings must be attributable to the example of the Social Science Association where a much greater level of women's involvement was evident from its foundation in 1857. Mary Carpenter, pioneer of the reformatory schools

---

94 *Ipswich Journal*, 22 August 1868.

95 *Devon Weekly Times*, 20 August 1869.

96 Susan D. Bernstein, '"Supposed differences": Lydia Becker and Victorian women's participation in the BAAS', in Clifford, Wadge, Warwick and Willis (eds), *Repositioning Victorian Sciences*, p. 86.

97 Rose Mary Crawshay, wife of Robert Crawshay of Cyfarthfa ironworks, Merthyr Tydfil, was a campaigner on educational issues and women's rights. Her paper on domestic service had been presented to and published by the Social Science Association in 1874. Angela V. John, 'Crawshay [née Yeates], Rose Mary (1828–1907)', *Oxford Dictionary of National Biography* (Oxford, 2004), accessed 29 September 2011.

98 *Daily News*, 28 August 1875.

Figure 4.4 Male and female delegates populate different sections of the Social Science congress at Manchester in 1879, The Graphic, 11 October 1879 (author's collection)

movement, addressed its inaugural meeting and went on to become the first woman to address the BAAS three years later. It is with some justification then that the NAPSS has been described as perhaps 'the first public platform for middle-class women in British history',[99] although here, too, there were parameters governed by the gender ideology of the day. Appropriate topics on which women spoke included reformatory and industrial schools, the employment of women, girls' education, infant mortality and the management of workhouses.[100] More politicised issues such as women's suffrage were off limits, and the jurisprudence and international law sections of the congress remained male dominated. A degree of segregation was introduced when the Bristol congress of 1869 was the first to incorporate a women's conference for the discussion of matters 'of special interest in relation to women's work'.[101] Women speakers at this event included Mary Carpenter, who addressed an evening meeting on the subject of female education in India, and Dr Elizabeth Blackwell who addressed the section on health. Contemporary reaction to this initiative was varied. The editors of the *Bristol Mercury* commented approvingly, saying that 'it is gratifying to think that the Bristol congress will always be memorable for having first enabled the fair sex to take that share in great work to which their sympathies, enthusiasm, and keenness of insight peculiarly entitle them'.[102] Other sections of the press were clearly suspicious of women-only meetings. The *Daily News* could not resist a jibe at the arrangements, complaining that, 'From their daily afternoon sittings

[99] For details see Goldman, *Science, Reform and Politics*, pp. 115–6.

[100] These details are given in *Glasgow Herald*, 8 September 1869.

[101] MCL, NAPSS, BR F306 S2, cutting from *Manchester Courier*, 25 September 1879.

[102] *Bristol Mercury*, 2 October 1869.

the members of the press even are to be excluded – a fact for which, considering that much talking means heavy notes, they are not altogether unthankful'.[103] On this occasion, arrangements for the women's conference were made in order to allow participants to attend the other sections of the meeting, and the female voice continued to be audible at the main sessions of the congresses. At Manchester in 1879 their presence in significant numbers at the key proceedings was noted, as was the fact that they 'made a material contribution to the debates of the day'.[104]

Each of the associations hoped, like the BAAS, to 'make the cultivators of Science acquainted with each other, to stimulate one another to new exertions, to bring the objects of science more before the public eye, and to take measures for advancing its interests and accelerating its progress'.[105] These were lofty ambitions and the extent to which they were achieved is very difficult to measure. Attendance figures could not reflect the extent to which local people benefited from or engaged with the knowledge being exchanged. The numbers who may have read lengthy newspaper accounts of the scientific business of the meetings can never be accurately quantified. Qualitative evidence demonstrates that those in attendance were sometimes uninspired by what they saw and heard. As a result, historians have questioned the degree to which the populations of the host towns were instilled with a sense of learning. RASE's mission, for example, was to spread knowledge of agricultural improvement for the benefit of farms in the vicinity of its annual meeting location. But Nicholas Goddard, in his history of the Royal Agricultural Society of England, expressed some doubt about its success in reaching 'the lower echelons of the agricultural community'.[106] The Social Science Association, as part of its mission to bring knowledge of the social sciences to the public, sought to encourage the participation of the working classes but, according to its historian, 'was ambivalent about their message'.[107] The RAI was founded with ideas of extending knowledge of ancient art, archaeology and antiquities among the wider public, but could not rid itself of the image of social exclusivity.[108]

But perhaps there is some justification for a more generous assessment of the ability of annual meetings to engage the populace of Britain's provincial towns. The initiatives of the associations and of local organising committees to incorporate free exhibitions, popular evening lectures and tailored events designed specifically to engage working people and women did succeed in widening participation to some elements of the meetings over the course of the study period. Further, a reconsideration of what constituted involvement or participation provides some scope to reassess the 'reach' of the annual meeting. To judge levels of engagement

---

103 *Daily News*, 30 September 1869.

104 MCL BR F306 F2, *Manchester Examiner*, 3 October 1879.

105 Bodleian, Dep. BAAS 1, Correspondence of John Phillips (foundation volume), David Brewster to John Phillips, 23 February 1831.

106 Goddard, *Harvests of Change*, p. 30.

107 Goldman, *Science, Reform and Politics*, p. 372.

108 Ebbatson, thesis, p. 12.

on the basis of evidence of interest in the ideas and knowledge imparted is to take a narrow measure. When viewed from a broader perspective as urban events, it is evident that annual meetings were not just about knowledge; novelty, celebrity, spectacle and excitement all formed part of the attraction of parliaments of science for people at all social levels of the community. The arrival of visitors and the appearance of street decorations were a source of curiosity, interest and sometimes pride in the host locality. These sentiments spanned across the social and gender divides and were expressed not just by those attending but by onlookers too. From this perspective it is possible to perceive a level at which annual meetings could be experienced by the wider populace but which has often fallen outside the purview of previous studies. Their inclusion in a more broadly defined notion of 'audience' adds a new dimension to our understanding of how parliaments of science were experienced by the urban populations of the provinces.

Chapter 5
Beyond the Meeting: Host Towns and the Parliaments of Science Effect

For all the excitement, spectacle and civic ritual generated by the scientific congresses of the nineteenth century, they were temporary events. At the end of meeting week, when delegates had left, decorations been removed and visiting newspaper correspondents departed, host towns could return to normality and restore to their original uses the buildings appropriated for the occasion. But the Victorian knowledge associations and the towns which hosted them were united in expecting annual meeting weeks to deliver benefits that lasted well beyond the time span of the meeting programme. In Exeter the mood immediately after the BAAS meeting of 1869 was buoyant: 'The British Association for the Advancement of Science has come and gone, leaving us all, we trust, the better for the visit. Contact with large, well stored minds, most of them free from narrow prejudices and stereotyped creeds is an excellent thing … We fancy we note the signs of improvement in the tone of thought all about us'.[1] As members of the Archaeological Institute departed Newcastle in September 1852, the editors of the *Newcastle Chronicle* reflected that, 'The amount of good effected by an association of this kind is not to be measured by the mere number of papers read or of places visited. The effects of mutual combination for a common object has a higher and more lasting influence …'[2]

While the belief that annual meetings would bring long-term benefits to their host towns was clear, less well defined was the exact form and duration of this 'lasting influence'. Historical studies of the associations and their annual congresses have hitherto done little to address this issue,[3] but the town-based perspective adopted here offers greater potential to assess the impact of these events on urban life in the provinces. In order to achieve this, evidence of the built environment, the cultural institutions and the activities of some of the key people involved in organising annual meetings is considered in this chapter. From an analysis of this data for a range of towns it is possible to detect a kind of parliaments of science 'effect' which, in the short term, galvanised towns into

1 *Devon Weekly Times*, 27 August 1869.

2 *Newcastle Chronicle*, 3 September 1852.

3 One exception to this is Lowe, 'The British Association and the Provincial Public', in MacLeod and Collins (eds), *The Parliament of Science*, pp. 118–44.

action over building projects and other urban improvements and, in the longer term, elevated cultural and educational functions to a new level of importance in town life. It is suggested that in many of the towns where scientific congresses were held, the increased awareness of the status and reputational gains to be made from being a centre of knowledge, even if only for a week, were not forgotten afterwards. The assets which had helped towns to attract the associations in the first place gained currency from their connection with a successful meeting. Key people, institutions and buildings, once praised by outsiders, became more readily valued by their own communities. Moreover, the involvement of local government leaders in the bidding and hosting processes fostered closer connections between municipal and cultural life. In some cases, the hosting of an annual meeting can be seen as part of a trajectory of urban development by which towns established themselves as more permanent knowledge centres and worked towards the establishment of civic colleges for the delivery of university education by the end of the nineteenth century.

'What Next?' was the question posed in one local newspaper as NAPSS delegates left Liverpool in 1858 at the end of their October meeting. 'The "Social Science" of the great manufacturing districts of Lancashire and Cheshire, and its outport on the Mersey', the same writer went on to suggest, 'can be studied only with advantage on the spot. If ever rightly learned it will be by and amongst men who live in the district'.[4] This idea that the work of social science should continue in Liverpool was shared by members of the local organising committee and by the Council of the NAPSS which, at its final meeting before the close of the congress, authorised the appointment of a local Liverpool Committee of the Association, with a grant of funds 'sufficient to meet its expenses during the current year'. Its aims were defined as 'those of the Association at large with the exception that they are confined to Liverpool' and a committee was appointed to direct its work, made up of representatives of the clerical and medical professions in the town as well as the merchant, J.T. Danson, and several men who had served on the local organising committee for the recent congress.[5] They set themselves a variety of tasks. The Reverend J.S. Howson, for example, carried out a report into the educational condition of the population of the town on behalf of a local education committee and in anticipation of the work of the commissioners for inquiring into the state of education in the country.[6] The initiative in Liverpool was followed in a number of other towns and, by 1865, branch associations in Bradford, Edinburgh and Aberdeen

---

4 NML MAL, John Towne Danson, D/D/III/17/5a, Newspaper Cutting Book, extract from *The Northern Times*, 19 October 1858.

5 NML MAL, John Towne Danson, D/D/III/15/3 Loose pamphlet and papers re. the National Association for the Promotion of Social Science, 'Address of the Liverpool Local Committee', February 1859.

6 NML MAL, John Towne Danson, D/D/III 8/3/3 Report to be made to the Council of the Association meeting at Bradford, 10 October 1858.

were listed alongside that of Liverpool in the annual *Transactions*.[7] After the initial enthusiasm of hosting a congress had worn off, such initiatives were difficult to sustain, and Goldman suggests that the Liverpool Local Committee was inactive by the mid-1860s.[8] Nevertheless, such ventures testified to the fact that local organisers and Association members believed in their own rhetoric and that far from being viewed as one-off events, the annual congresses were conceived as a launch-pad for more lasting knowledge promotion.

In the 1850s and 1860s it was particularly in the realm of sanitary improvement that town leaders hoped to benefit from the hosting of learned meetings. In 1852, ahead of the British Association meeting in Hull, the editors of the *Hull Packet* reminded their readers that 'the sanitary condition of towns has received frequent consideration' from the BAAS, and urged members of the Local Board of Health and to 'take the opinion of the Association' on public health-related matters.[9] In Leeds, meanwhile, local journalists were even more direct in making the link between scientific knowledge and sanitary reform: 'The spread of scientific tastes, the thirst for knowledge, the ambition to acquire information on those subjects discussed by the British Association, would perhaps be the best guarantee that Leeds was not only sincere, but likely to be successful in its campaign against the filth, poverty, crime, disease and other horrors that revel in its lowest suburbs'.[10] It was the work of the Social Science Association which had a more direct bearing on debates over urban sanitation, although the extent to which the NAPSS was able to instigate legislation specifically designed to tackle these problems has been questioned.[11] Nevertheless, there were some successes. George Hastings noted with satisfaction 'the great impetus given to the Drinking Fountain movement by our Liverpool meeting'. As a result of a report on the impact of fountains there on reducing drunkenness, 'the idea has spread through the whole kingdom and now there is hardly a large town in which it has not been more or less carried out'.[12] Hastings also claimed credit for the Association in the development of the Scotland Police Bill which, he informed Lord Brougham, 'was passed chiefly through the exertions of Mr Lindsay, Provost of Leith, a member of the Association, who learned at our meeting the importance of sanitary measures'.[13] After the Social Science congress at Glasgow in 1860, where Edwin Chadwick presided over the

---

7 *Transactions of the National Association for the Promotion of Social Science, Sheffield Meeting* (London, 1866), p. xx.

8 Goldman, *Science, Reform and Politics*, p. 86.

9 *Hull Packet*, 10 September 1852.

10 *The Leeds Mercury*, 25 September 1858.

11 Ronald K. Huch, 'The National Association for the Promotion of Social Science: its contribution to Victorian health reform, 1857–1886', *Albion*, 17/3 (1985): 279–99.

12 UCL Special Collections, B MSS, 7646, George W. Hastings to Lord Brougham, 1 September 1859.

13 UCL Special Collections, B MSS, 8790, George W. Hastings to Lord Brougham, 6 September 1863.

public health section, Provost Lindsay composed a circular letter which was sent to provosts of burghs and chairmen of commissioners of police throughout Scotland. In it, he asserted that most large towns in Scotland were 'suffering in a sanitary aspect, for want of sufficient powers in the authorities to accomplish what is needful for the public health'. Rather than applying for numerous local acts to address these problems as and when they arose, he urged them to consider the merits of a general police act which would 'embrace all the powers contained in the existing English and Scotch Acts, with such others as might be suggested by the local authorities of the burghs'.[14] As well as contributing to the passage of the Police and Improvement (Scotland) Act, 1862, it also raised Lindsay's profile in reforming circles nationally, with one contemporary describing him in a meeting of the inhabitants of Leith, ahead of the Social Science congress in Edinburgh in 1863, as someone who was 'looked upon as one of the greatest pioneers of sanitary reform'.[15]

Social Science congresses, where public health, sanitation and other issues of great significance to towns were debated, provided talented local men with a platform to share and extend their knowledge of these subjects. The expertise they gained helped some to extend their influence in public life. In Bristol, the shipping merchant Mark Whitwill enhanced his local reputation, built up through his chairmanship of the Chamber of Commerce, by participating in annual Social Science congresses, both in Bristol in 1869 and elsewhere in the country. He attended NAPSS annual meetings at Glasgow and Cheltenham, preparing a written report on the former in which he summarised the major contributions on education, crime, public health and the social condition of the working population.[16] His developing interest in these issues stood him in good stead for a career in local government. He entered local politics in Bristol in the Liberal interest, serving on the School Board from 1871, and was first nominated an alderman in 1883.[17] In other towns, BAAS meetings provided a similar bridge for local men to acquire a national reputation or to cement a career in local government. In the year following the Nottingham meeting in 1866, the town's former mayor, William Felkin, who played a central role in persuading his townsmen of the merits of hosting a meeting, completed and published the book which established his reputation as a leading authority on the statistics and history of the industry in which he had made his living. His *History of the Machine-Wrought Hosiery and Lace Manufactures* was published in 1867 with support from a distinguished list of subscribers including Sir Roderick Murchison.[18] Likewise, it was probably no coincidence that the Nottingham antiquary, Thomas Close, who served as a vice-

---

14 A copy of this letter was published in *The Caledonian Mercury*, 18 January 1861.

15 Comments of Professor Archer, quoted in *The Caledonian Mercury*, 17 June 1863.

16 BRO 38605/X/5, The Bristol Incorporated Chamber of Commerce and Shipping: Mark Whitwill's Report of the Social Science congress held at Glasgow on 30 September 1874.

17 Alfred B. Beavan, *Bristol Lists: Municipal and Miscellaneous* (Bristol, 1899), p. 83.

18 William Felkin, *A History of the Machine-Wrought Hosiery and Lace Manufactures* (London, 1867), p. vii–xii.

president of the Nottingham BAAS meeting, published a volume on *St Mary's Church Nottingham: its probable architect and benefactors* in the same year as the BAAS visit. The *Nottinghamshire Guardian* described Close as 'our intelligent townsman' in naming him, along with the distinguished BAAS scientists who were due to assemble in the town for the annual meeting.[19]

The opportunity to mix with men of national and international scientific renown was a great boost to members of local or regional institutions, bringing their own work to wider public attention as well as raising their status in the immediate locality as they rubbed shoulders with distinguished visitors. In Manchester, the BAAS meeting of 1842 was an occasion to celebrate the achievements of the town's established scientific figures, particularly the chemist and natural philosopher John Dalton, whose image appeared on the ticket issued to British Association delegates attending the meeting.[20] It also propelled a new generation of local scientists into the spotlight.[21] Dalton's pupil, James Joule, the son of a brewer from Salford, addressed the Chemistry Section at the meeting on the subject of the electric origin of the heat of combustion, while the engineer William Fairbairn delivered a paper to the Mechanics Section on coal combustion and the prevention of smoke.[22] Both had previously given lectures at the Royal Institution in Manchester, but went on to contribute to later British Association meetings, Fairbairn acting as BAAS president in 1861 and Joule chosen to preside in 1873 and in 1887, but prevented by ill-health.[23] In Swansea, William Robert Grove, already elected a fellow of the Royal Society and appointed professor of experimental philosophy at the London Institution,[24] distinguished himself with the British Association, first in persuading them to hold their annual meeting of 1848 in the town, and then in his role as vice-president of the Chemical Science section during the Swansea meeting.[25] He went on to hold the presidency of the

---

19 *Nottinghamshire Guardian*, reprinted in *Daily News*, 17 August 1866.

20 Dalton had been a stalwart of the Manchester Literary and Philosophical Society, but his work on atomic weight theory brought him to a much wider notice. Frank Greenaway, 'Dalton, John (1766–1844)', *Oxford Dictionary of National Biography* (Oxford, 2004; online ed., Oct. 2006), accessed 9 March 2012.

21 Robert H. Kargon, *Science in Victorian Manchester. Enterprise and Expertise* (Manchester, 1977), pp. 32–3.

22 *Report of the Twelfth Meeting of the British Association for the Advancement of Science held at Manchester in June 1842* (London, 1843); Fairbrother, Birks, Wolfe Mays and Morgan, 'The history of science in Manchester', in Carter (ed.), *Manchester and its Region*, pp. 192–5.

23 Burnley, 'Fairbairn, Sir William'; Crosbie Smith, 'Joule, James Prescott (1818–1889), *Oxford Dictionary of National Biography* (Oxford, online edn. 2011).

24 See M.L. Cooper and V.D.L. Hall, 'William Robert Grove and the London Institution', *Annals of Science*, 39 (1982): 229–54.

25 *Report of the Eighteenth Meeting of the British Association for the Advancement of Science* (London, 1849), p. xv.

British Association in 1866 when the annual meeting was held in Nottingham. In Birmingham in 1865, the scientific significance of the work of James Chance, local glass manufacturer, was praised by Professor W.A. Miller in his presidential address to the Chemistry Section.[26] Chance, already well known in some scientific circles for his work in developing glass for use in lighthouse lenses, went on to exhibit his designs at the Paris Exhibition of 1867 and to address the Institution of Civil Engineers on aspects of his lighthouse work.[27]

As well as delivering a significant fillip to the work of local scientists who might otherwise have struggled to gain national recognition, parliaments of science could also provide a platform for entry into the highest municipal office. British Association meetings sometimes had the effect of propelling learned townsmen into mayoral office. In Hull, Dr Henry Cooper, one of the principal local organisers of the 1853 BAAS meeting in the town, was elected to serve as mayor for two consecutive terms from 1853–55. Already distinguished in the town as physician to the Hull Infirmary and as a sanitary reformer, he presented papers to the Association's Statistics Section on 'the mortality of Hull in the Autumn of 1849' and on 'the prevalence of diseases in Hull'.[28] Likewise in Exeter, silversmith Henry Ellis was inaugurated as mayor in November 1868, so that he was in office for the year of the British Association visit. His reputation had been built up through his involvement in county scientific circles and in particular his active membership of the Devonshire Association for the Advancement of Science, founded in 1862.[29] He went on to attend BAAS meetings in Nottingham in 1866 and Norwich in 1868 as part of a delegation to press the case for an annual meeting in Exeter. To have such men ensconced in the mayoral office in the year of an important scientific congress sent out a powerful message about the interconnection between scientific and municipal life in a community and its impact was not lost on distinguished association members. At Newcastle upon Tyne, when the British Association visited in 1863, the town's mayor, Isaac Lowthian Bell, addressed the Chemical Section with a paper on the manufacture of aluminium,[30] attracting plaudits from Sir Roderick Murchison, Professor Alex Williamson and others for his 'distinguished position as a scientific man'.[31]

---

[26] *Report of the Thirty-Fifth Meeting of the British Association for the Advancement of Science, held at Birmingham in September 1865* (London, 1866), p. 23.

[27] Welch, 'Chance, Sir James Timmins'.

[28] 'Modern Hull', in Keith J. Allison (ed.), *A History of the County of York East Riding: Volume 1: The City of Kingston upon Hull* (London, 1969), pp. 215–86; *Report of the Twenty-Third Meeting of the British Association for the Advancement of Science held at Hull in September 1853* (London, 1854), p. x.

[29] *Report of the Second Meeting of the Devonshire Association for the Advancement of Science Literature and Art, held at Plymouth, July 1863* (London, 1864).

[30] *Report of the Thirty-Third Meeting of the British Association for the Advancement of Science, held at Newcastle Upon Tyne In August and September 1863* (London, 1864), p. viii.

[31] *Leeds Mercury*, 3 September 1863. See also Campbell, 'Men of science in nineteenth century Newcastle', p. 60.

The reputations of these talented provincial men of science did not rest on the hosting of scientific meetings in their home towns, but their prominence at these events gave the host locality an added scientific credibility, enabling them to claim that they had expertise to impart as well as much to learn during the course of the annual meeting. Such was the eminence of men like Isaac Lowthian Bell in Newcastle in 1863 that the dynamic of the meeting was altered, according to one commentator, from an occasion where visiting experts came to share their knowledge with locals, to one where 'the visitors came to sit at the feet of Newcastle men who were the acknowledged leaders in their respective fields'.[32] But there was more to be gained by participant towns than simply words of praise and congratulation from visiting experts for their scientific attainments. By contributing to public debates on social science and sanitary reform, and by sharing their expertise in branches of Chemistry and Mechanics, annual meetings tied the provincial host towns into a wider project of national enlightenment through scientific and technological achievement.[33] This was arguably becoming more important by the last quarter of the nineteenth century. Events like the Paris exhibition, attended by James Chance, revealed the advancing state of European science and engineering and heightened the sense of scientific prowess as a sign of national vitality. Scientific discovery was less about self-improvement, as in earlier eras, and more about the progress of society and the nation.[34]

The boost given to the standing of talented individuals through involvement in this national science project was matched by a discernible pattern of stimulus to local scientific and literary societies of various descriptions in the immediate aftermath of the visit of one of the major national knowledge associations. As the editors of the *Newcastle Courant* put it, as they reflected on the British Association meeting in Liverpool in the knowledge that their town would be the host a year later, 'all liberally-founded public institutions for the dissemination of knowledge receive an impetus from the presence of learned strangers from so many quarters; their local supporters are encouraged to persevere in their endeavours to overcome every difficulty, for the sake of the great good which the societies produce'.[35] It was a view shared by the Associations, which saw themselves as energising the societies operating in the localities where their meetings were held. The Royal Archaeological Institute, for instance, claimed that its 'visits, attended as they are by Archaeologists and eminent men from all parts of the kingdom, have given a great stimulus to the local taste for Archaeology, and have, in consequence, like those of the Royal Agricultural and other central bodies, been favourably regarded

32 Campbell, 'Men of science', p. 60.

33 Carol E. Harrison and Ann Johnson, 'Introduction: Science and National Identity', *Osiris*, 24/1, *Science and National Identity* (2009): 1–14.

34 Jordanova, 'Science and Nationhood', pp. 197–200; Frank M. Turner, 'Public Science in Britain, 1880–1919', *Isis*, 71/4 (1980): 592.

35 *Newcastle Courant*, 22 September 1837.

by local societies'.[36] Exactly what form this benefit might take, however, was less apparent. In some cases, existing societies received a funding injection from the reserves of local organising committees. Such was the case in Bristol after the British Association visit of 1836 when the local committee gave grants of £50 to the Bristol Philosophical Society, £80 to Clifton Observatory and £20 to Bristol Mechanics Institute (a sum which was subsequently increased to £25).[37] In Newcastle in 1863, a healthy surplus of over £350 in the subscribers' fund after the British Association meeting of that year allowed the local organising committee to make a grant of £200 to the town's Natural History Society, with the remainder divided between the Mechanics Institute's building fund and the local Fine Arts Society.[38]

Encouraging as these developments were, in the long run they did not always live up to the hyperbolic predictions of local journalists and guidebook writers. Committee members of the Hull Literary and Philosophical Society, meeting in May 1854 to reflect on the events of the preceding year, agreed that the BAAS visit formed 'a memorable epoch' in their history, and one which 'connected the society with the highest class of the cultivators of science'. But at the same time they worried that enthusiasm was not being sustained, noting that attendance by non-members of the society at recent lectures had been 'extremely small' and that even society members' attendance had been disappointing. In a review of membership figures it was also noted that the steep rise from 286 in 1852 to 351 in 1853, the year of the BAAS meeting, had also slowed considerably to reach 361 members.[39] Similarly, the Chester Archaeological Society, though buoyed by the meeting of the Royal Archaeological Institute in the town in 1857 when its membership reached 177, struggled to maintain interest in subsequent years with no minutes kept of meetings between 1859 and 1864.[40] Rather like the effects of a religious revival on levels of spiritual devotion, it seemed that the annual meeting could not sustain, indefinitely, heightened levels of scientific participation. But where a more substantial financial investment could be made in the opening of new buildings, a more permanent mark was made on communities in the form of new premises occupying prominent positions in the Victorian streetscape.

In Chester, the need of both the local archaeology and natural history societies to find a permanent building in which to house their growing collections led to a joint venture to build a museum premises in the town's Grosvenor Street.[41] Inaugurated with the launching of an appeal fund in November 1883 to which

---

36 *The Archaeological Journal*, 27 (1870): 2.

37 BRO, 32079, Minute Book of the Reception Committee for the British Association, 19 September 1836.

38 *Newcastle Courant*, 13 November 1863.

39 *Hull Packet and East Riding Times*, 12 May 1854.

40 Alan G. Crosby, *The Chester Archaeological Society. The First One Hundred and Fifty Years, 1849–1999* (Chester, 1999), p. 20.

41 Glenys Lloyd-Morgan, 'The early growth of archaeology in Chester: 1849–1895', *Journal of the Chester Archaeological Society*, 72 (for 1992/3): 15–23.

the Duke of Westminster contributed £4,000 and the gift of the site, the building also housed the Chester School of Art and the town's science classes.[42] As such it provided an important new scientific and educational hub not just for the town but for the wider region as its full name, the Grosvenor Museum for Chester and North Wales, indicated. The building was formally opened in a civic ceremony on 10 August 1886, a day before the commencement of the Royal Archaeological Institute's annual meeting in the town and was therefore timed to attract maximum attention as well as to impress the distinguished visitors in town for the congress. Likewise in Hull, members of the Literary and Philosophical Society, who had played a key part in persuading the British Association to visit the town in 1853, used the momentum of the visit to press ahead with plans for a new building in which to house their own collections along with the town's subscription library. Their new Royal Institution, opened by the Queen in October 1854, gave a physical permanence to the place of scientific learning in the town and seemed to lend some credence to the excited claims of the editors of the *Hull Packet* in the immediate aftermath of the meeting, that 'the inhabitants of Hull have given substantial proof of their title to be considered as something better than a society of mere money-getters'.[43] In a new town guidebook published later in the century, the Royal Institution building was given a detailed description, accompanied by a double-page illustration.[44]

The ability to furnish lecture spaces and museum accommodation was not just a practical consideration which, as we saw in Chapter 2, enabled towns to present themselves as credible meeting hosts. These places were also important signifiers of a town's cultural identity and civility and, with visitors arriving ready to compare what they saw with facilities in their own towns, it was worth concentrating resources on making these premises as impressive and extensive as possible. In Liverpool in 1837, for example, the corporation gifted land for a new mechanics' institution, and public subscriptions were raised for a building with a central lecture theatre capable of accommodating 2,000 people, specifically with the visit of the British Association to the town that year in mind. Despite a serious fire in April which caused damage estimated at £3,000 to the nearly completed structure, repairs and renovations were still finished in time for the arrival of the BAAS.[45] The holding of the formal opening of the building during the meeting week meant that a number of the visiting scientists and dignitaries could attend the event, including the Earl of Burlington, the Association's president.[46] As the author of one study of the effects of BAAS meetings has claimed, the prospect of a forthcoming annual

42 *Morning Post*, 10 August 1886.

43 *Hull Packet and East Riding Times*, 16 September 1853.

44 William Shepherdson, *The New Hull Guide: a modern picture of Hull and Visitor's handbook to the various objects of interest in Hull and neighbourhood* (Hull, 1875), pp. 128–9.

45 For details of the fire damage, see *Morning Post*, 3 April 1837.

46 *Liverpool Mercury*, 22 September 1837.

meeting was enough to galvanise towns into action and motivate them to expend effort and money in bringing their scientific facilities up to scratch.[47]

The improving effect on the urban fabric of towns where meetings were hosted was not confined to buildings associated with the local scientific societies and mechanics institutes. The availability of a large hall, as we saw in Chapter 2, was almost an essential prerequisite for the hosting of a major annual meeting. Although such buildings were being constructed in towns large and small all over urban Britain in the Victorian and Edwardian period, the parliaments of science effect hastened efforts in some places. In Leeds, the town hall building project, ongoing since 1853, was brought to a successful conclusion in time for the BAAS visit in 1858. It was opened as a reception room during the meeting week and housed the sectional meetings as well as evening events including a conversazione.[48] In Bradford, the opening of the new town hall just days before the arrival of British Association delegates for their annual meeting in 1873 was similarly arranged with the congress in mind. Although constructed to provide more adequate accommodation for the municipal offices of the town, the new hall was part of a series of improvements to the built environment which included a new Exchange, Mechanics' Institute and Free Library which 'quite transformed the appearance of the grimy town' to the extent that, 'Those who knew Bradford twenty years ago, and have not visited the town in the meantime, would have difficulty in recognising it in 1873'.[49] The celebrations to mark the opening of the hall in mid-September created a 'high festival mood' just as the British Association delegates were about to descend, and contributed to a general air of optimism about the prospects for the forthcoming meeting.[50] In Exeter, too, news that the British Association would be visiting in 1869 prompted townsmen to address the issue of public rooms. After discussing various options for enlarging existing premises to meet the needs of the meeting,[51] a more ambitious decision was taken to build a new hall, funded privately by shareholder contributions. The new Victoria Hall, located on the town's Queen Street, was completed in time to accommodate the estimated 2,000 people who attended the presidential address by Professor Stokes to mark the opening of the meeting,[52] and remained the largest public hall in the town until its destruction by fire in 1919. Although they were later put to more general purposes, these buildings were closely identified with the parliaments of science which marked their inauguration, and provide a reminder of the physical mark such events could leave on the townscape.

This zeal for improvement extended not just to buildings which might help accommodate delegates during the annual meeting. The prospect of a congress

---

47 Lowe, 'British Association', p. 125.

48 *Leeds Mercury*, 16 September 1858.

49 *Leeds Mercury*, 21 January 1873.

50 See *Leeds Mercury*, 10 September 1873 and 17 September 1873.

51 *Devon Weekly Times*, 11 September 1868.

52 *Devon Weekly Times*, 20 August 1869.

could also have the effect of incentivising local government leaders to complete more symbolic projects in time for the occasion. In Manchester in 1866, plans to commemorate Prince Albert with the construction of a fitting memorial were hastened by the approach of the Social Science congress in the autumn of that year. The Prince Consort Memorial Fund committee, meeting in the town in April,

> ... determined to invite contracts at once for the completion of the pedestal, the remainder of the carving and the grille round the monument. Great confidence was expressed that by these measures the memorial would be completed by the month of October next, previous to the first meeting of the Social Science Association which will be held in this city about that time.[53]

In the event, the October deadline was missed and the memorial was inaugurated in January 1867,[54] but the effect of the congress in accelerating the process was unmistakable. In other towns too, the timing of efforts to secure a permanent memorial to the late Prince Consort were affected by the approach of an annual meeting. In Dundee, the opening of the Albert Institute, a new building equipped with museum, gallery and lecture space occupying a prominent site close to the centre of the town, followed a similar pattern. Although discussions over an appropriate memorial to Prince Albert predated Dundee's invitation to the British Association, the Institute's development was closely bound up with its dealings with the BAAS. Delegation members returning from the annual meeting in Newcastle in 1863 reported that 'it would be desirable to have an educational institute of some sort or other established'[55] and, once work on the new institute was underway, the decision was taken to postpone the invitation to the BAAS until 1867 so that it would be complete and could be fully utilised during the meeting.[56] It was a similar story in Exeter where a public subscription for funds to build an Albert Memorial Museum had been underway since just after the Prince's death in 1861, but the BAAS visit in 1869 provided the final push needed to complete the work and open the new institution to the public.[57] Albert's role as a patron of science and the arts, and his regular attendance at British Association meetings, made it incumbent on towns hosting learned meetings, especially in the years immediately following his death in the 1860s, to have some fitting commemoration in place which would meet the approval of visitors.

It was not just town authorities and members of local societies who could be found casting a critical eye over the townscape in the run up to hosting an

---

53 MCL, BR F306 N1: National Association for the Promotion of Social Science, cutting from *Manchester Examiner and Times*, 11 April 1866.

54 See report on the inauguration in *Manchester Times*, 26 January 1867.

55 *Dundee Courier and Argus*, 5 February 1864.

56 *Dundee Courier and Argus*, 3 September 1867.

57 Louise Miskell, 'Meeting Places: the scientific congress and the host town in the south west of England, 1836–1877', *Urban History*, 39/2 (2012): 255–6.

annual meeting. Their efforts were matched by private inhabitants who, prior to the arrival of meeting delegates, carried out repairs and improvements to their premises which were designed to last well beyond the duration of the meeting. In Leicester, before the arrival of the Royal Agricultural Society of England for its annual show in 1868, there was a flurry of improvement activity around the town as householders and traders strove to make their properties look as modern and attractive as possible. According to the *Leicester Chronicle*:

> … architects and builders, painters, paper-hangers, and white-washers, have been in constant request for months past and must have had a fine time of it. Antiquated shop-fronts have given place to ornate designs in stone, encasing huge panes of plate glass; while the anxiety displayed amongst the occupants of private houses to modernize their dwellings, must now and then have caused architects some trouble, and required much forbearance. So great has been the rage for bay windows, that the wonder is that any others remain which the engagements of builders or the purse of occupiers did not prevent from being converted. Many of the old windows have been thus replaced with great advantage to the occupiers, doubtless; but why on earth some have been so transformed, it would puzzle a philosopher to tell.[58]

In contrast to the other annual meetings, RASE shows left their mark in a rather different way on the towns in which they were held. As RASE finance committee members were quick to point out to local authorities, the work involved in draining, levelling and fencing land, and sometimes improving access roads to the showground, could enhance the value of the land on which the event was staged.[59] But there was no simple answer to the question of what should happen to this land after the show. In some towns, like Newcastle upon Tyne, the area occupied by the RASE showground was open access land which authorities were keen to preserve from any kind of development. The Newcastle town moor where the RASE show was held in 1846 and 1863 had been used for centuries as the site of town fairs, races, volunteer meetings and for the playing of cricket, football and bowling.[60] Local council members resisted all suggestions for its development, even for recreational purposes as a more formal park. However, the legacy of the agricultural shows found its way into legislation by 1870 when the Newcastle upon Tyne Improvement Act included a clause which allowed for the 'appropriation of part or parts of the town moor for the space of not exceeding ten continuous days

---

[58] *Leicester Chronicle*, 11 July 1868.

[59] See for example, MERL, SR RASE BII/4 Minutes of Finance Committee, 1870–87, 5 November 1879.

[60] Robert J. Charleton, *A History of Newcastle-on-Tyne. From the Earliest Records to its Formation as a City* (Newcastle, 1885), pp. 383–4.

for the purpose of agricultural shows, public reviews or other public purposes'.[61] Ironically, the movement for parks and open spaces which gathered pace in the second half of the nineteenth century did not necessarily improve the capacity of towns to accommodate agricultural society show grounds. In Leicester, for example, a group of town councillors voiced strong opposition to the idea that the town's Victoria Park should be offered to RASE for its 1894 show. The rights of cricketers to access their playing pitch, children to use the playground and walkers to access the paths and walkways would, they argued, all be compromised by the closing off of the park for the duration of the show.[62]

In Cardiff, by contrast, the use of private land owned by the Marquis of Bute at Cathays Park for the showground of the Royal Agricultural Society in 1872 helped draw attention to the attractions of this site of over 40 acres so close to the town centre. Visiting newspaper correspondents in town for the show wrote admiringly of 'his beautiful park', 'a noble expanse of ground' and 'a convenient and beautiful site'.[63] Whether or not such sentiments had a direct influence on how Cathays Park was viewed in Cardiff is unclear, but two years later when discussions were underway about the need for an Improvement Act for the town, a number of members of the corporation raised the issue of Cathays, arguing that the site should be acquired for the town for development as a public park.[64] Although the prohibitive costs of this delayed any such action, the area continued to be used with the consent of the Marquis as a venue for shows and events, including the Glamorgan Horticultural Society show in August and the South Wales National Eisteddfod in September of 1879. Not for another eight years did its development as a planned civic quarter for Cardiff begin, with the grant of a portion of the land for the location of the University College of South Wales and Monmouthshire in 1887.[65]

If RASE visits ignited debates in towns about the use of open spaces, so the meetings of the Royal Archaeological Institute raised awareness of the fate of historic sites, especially in towns where rapid demographic and commercial growth had resulted in the removal of old buildings. In Hull, the *Visitors' Descriptive Guidebook*, published to mark the Royal Archaeological Institute's meeting there in 1867, noted that the growth of the town as a commercial trading port had come at the expense of some of its most significant historic places:

---

61 E.M. Halcrow, 'The Town Moor of Newcastle upon Tyne', *Archaeologia Aeliana*, 4th series, 31 (1953): 161.

62 'Leicester Town Council. The Royal Show and the Victoria Park', *Leicester Chronicle and Leicestershire Mercury*, 21 July 1894.

63 See for example reports in *The Standard*, 17 July 1872 and *Jackson's Oxford Journal*, 13 July 1872.

64 See editorials and reports of Cardiff Corporation meetings in the *Western Mail*, 9 October 1874 and 13 November 1874.

65 *Western Mail*, 14 April 1887. For a full account of the later development of Cathays Park see Ian Morley, 'Representing a city and a nation: Wales' matchless civic centre', *Welsh History Review*, 24/3 (2009): 56–81.

> There is very little left of Old Hull; there are little or no remains of the venerable edifices which once afforded shelter to the Augustine, Carmelite and Carthusian orders; the last relic that would have excited the curiosity of the antiquarian, namely the ancient Citadel, has disappeared to make room for the increasing commerce of one of the principal Seaports in the British Empire.[66]

It was the same story in other towns and cities, not least London, where major engineering projects directed by the Metropolitan Board of Works from 1855 resulted in the removal of prominent landmarks like Northumberland House and Temple Bar.[67] But in parallel, and perhaps partly in response to this trend, there was a growing popular interest in historic sites and monuments and the development of a kind of historical tourism from the 1840s.[68] Interest in castles and country houses as places to visit, combined with the activities of local antiquarian and archaeological societies, some 56 of which were founded in England alone between 1836 and 1886,[69] meant that more attention was focused on the condition and significance of local historic sites. Archaeological Institute meetings, which paid so much attention to the study of local archaeological and historic places, were sometimes the stimulus to the identification and preservation of monuments which otherwise risked falling into decay. In Lincoln in 1848, for example, attention focused on the condition of sites of interest which visitors might want to inspect during their stay. The local Topographical Society donated money towards the clearing out of a Roman tunnel in the roof cellar of one of the town's old inns, and towards the cleaning of a building known as the 'Jews House'. In both cases, it was hoped that the RAI would match their financial input with additional donations.[70] Once underway, the meeting drew attention to the condition of some important local sites, in particular of the ancient parish Church at Stow.[71] This important Anglo-Saxon building had been the focus of interest for members of the Lincolnshire Society for the Encouragement of Ecclesiastical Architecture since its formation in the early 1840s,[72] but it was the excursion of RAI members during

---

66 John Symons, *The Visitors' Descriptive Guide Book to Kingston upon Hull* (Hull and London, 1867), p. 2.

67 John Earl, 'London Government: a record of custodianship', in Michael Hunter (ed.), *Preserving the Past. The Rise of Heritage in Modern Britain* (Stroud, 1996), pp. 57–76.

68 Peter Mandler, '"The Wand of Fancy": the historical imagination of the Victorian Tourist', in Marius Kwint, Christopher Breward and Jeremy Aynsley (eds), *Material Memories. Design and Evocation* (Oxford, 1999), pp. 127–9.

69 Ebbatson, 'Context and Discourse', in Vyner (ed.), *Building on the Past*, p. 26.

70 Birch, 'The "archaeologians" visit Lincoln', p. 170.

71 Hill, *Victorian Lincoln*, p. 80.

72 Mark Spurrell (ed.), *Stow Church Restored, 1846–1866. Lincolnshire Record Society Publications*, 75 (Suffolk, 1984), p. xxi.

their meeting in Lincoln in 1848 which drew wider attention to its neglected state and sparked renewed efforts to raise funds for its renovation.[73]

Manufacturing towns, though they may have lost much of their archaeological heritage in the face of commercial and industrial development, found different ways of packaging and preserving specialised local knowledge showcased during annual meetings. At the BAAS meeting in Newcastle in 1863, where the presidency was held by Sir William Armstrong, the armaments manufacturer and industrialist of the town, the opportunity was seized to use the occasion to promote the industrial and commercial assets of the area and its part in the progress of railways, mining, armament manufacture and other industries. This was a key theme of Armstrong's inaugural address at the opening of the conference, but it was also taken up and developed in a volume on *The Industrial Resources of the Tyne, Wear and Tees*, published after the meeting. The publisher's preface noted that 'the visit of the British Association was deemed a fitting occasion' on which to present information on 'the introduction into this district of various manufactures for which it is now so celebrated [and] the numerous improvements which had their origin here'.[74] The volume contained illustrations of apparatus and machinery developed by Armstrong and other manufacturers at their works, making it an extended advertising document for the achievements and expertise of Newcastle's principal industrialists.

The publicity potential of the Newcastle volume was not lost on organising committees for BAAS meetings in other manufacturing centres. At Birmingham, two years later, a local industries committee was assembled, headed by the local historian Samuel Timmins, to organise a series of papers for presentation to the Economics and Social Science section of the congress, and for subsequent publication. The resulting volume contained essays on the Birmingham metal trades and also a host of other manufactures from rope-making to umbrella manufacture and coffin-making.[75] A reviewer in one local newspaper predicted that the effect would be to bring greater knowledge of 'the infinitely varied character of its manufactures' and 'the wonderful ingenuity of its skilled workmen' to a wider audience and suggested that towns like Leeds, Manchester and Sheffield should follow the example of Birmingham and Newcastle.[76] From the point of view of the associations, such publications represented a pleasing outcome from their annual meetings, providing enduring evidence that a valuable knowledge-gathering exercise had taken place. This was certainly the case in the NAPSS

---

73 Birch, 'The "Archaeologians" visit Lincoln', p. 172.

74 William G. Armstrong, Isaac Lowthian Bell, J. Taylor and Dr Richardson (eds), *The Industrial Resources of the District of the Three Northern Rivers, the Tyne, Wear and Tees* (London, 2nd edition, 1864), p. iii.

75 First published as Samuel Timmins (ed.), *The Resources, Products and Industrial history of Birmingham and the Midlands Hardware District* (London, 1866). New impress, S. Timmins (ed.), *Birmingham and the Midland Hardware District* (Frank Cass, 1967).

76 *Birmingham Daily Post*, 3 September 1866.

when news filtered through of the publication of a volume of studies of strikes and trade societies, based on reports collected for the Glasgow meeting in 1860. It gave George Hastings particular cause for satisfaction that these were 'materials which would not otherwise have been brought together'.[77]

It was not just existing areas of expertise which received a boost from annual meetings. The debates and discussions of association delegates could also stimulate the creation of new institutions in some locations. A flurry of new societies appeared in Bradford in the years following the British Association meeting of 1873, including the West Riding Photographic Society, established in 1874, and the Bradford Naturalists' Society, formed a year later.[78] It seemed that the British Association 'effect' also extended to the nearby town of Halifax which established a new scientific society in 1874.[79] In Dundee, the British Association visit of 1867 also left an institutional legacy in the town with the formation of a Natural History Society the same year and a Naturalists' Society in 1874. The latter attracted a membership of over 400 in just three years, making it the largest society of its kind in Scotland, thanks in part to its popular evening exhibitions and conversaziones, many of which were held in the Albert Institute building.[80] In Norwich, a Meteorological Society was founded after the meeting of the British Association there in 1868.[81] The subject of meteorology had featured prominently in the business of the BAAS that year, not least because a number of scientists had witnessed a meteor shower in August. A report on this and other recent meteor observations by a committee from the Royal Observatory, Greenwich, was published in the annual meeting report, as was a brief synopsis of a paper by George Forbes outlining his observations of the meteor shower which he presented to the Astronomy Section.[82]

There were many other examples of town-based educational developments which gained momentum from the visit of a national association. New initiatives sometimes germinated at annual congresses, such as at Norwich where discussion of female education at the Social Science congress held in the town in 1873 marked the commencement of a local campaign to improve facilities for the schooling of local girls. With support from the Girls' Public Day School Company, a High School

---

77 *Trades' Societies and Strikes. Report of the Committee on Trades' Societies appointed by the National Association for the Promotion of Social Science* (London, 1860); UCL Special Collections, B MSS, 7668, George W. Hastings to Lord Brougham, 15 December 1860.

78 For a list of institutions of literature, science and art in Bradford, see *Post Office Bradford Directory, 1879–80* (Bradford, 1879), p. 37.

79 Howarth, *The British Association for the Advancement of Science: A Retrospect*, p. 96.

80 Finnegan, 'Natural History Societies in late Victorian Scotland', p. 61.

81 Lowe, 'The British Association and the Provincial Public', in MacLeod and Collins (eds), *The Parliament of Science*, p. 125.

82 *Report of the Thirty-Eighth Meeting of the British Association for the Advancement of Science, held at Norwich in August 1868* (London, 1869), p. 344, p. 13.

for girls was opened in the town in the spring of 1875. Similar initiatives followed in other towns but considerable satisfaction was felt in Norwich for having been in the first flush of developments on this front. In a new gazetteer and directory, published eight years later, a paragraph was given over to explaining that the idea for the school had 'originated with the Social Science Congress of 1873' and that it was one of the oldest of the Girls' Public Day School Company institutions.[83] Elsewhere it was the acceleration of initiatives which might otherwise have been long delayed that was notable. In Birmingham in 1865, a new Free Library was opened by the mayor on the first day of the British Association's annual meeting in the town. As one local newspaper reporter pointed out, the opening of the library, 'though not connected with the visit, yet associated with it … will be a pleasing introduction to the events of the week'.[84] The entertainment laid on by the mayor was followed by Lord Stanley's address to mark the opening of the congress and the new library provided accommodation for some of the business of the Geology Section during the ensuing week. Ratepayer opposition was a barrier to the adoption of the Public Libraries Act of 1855 in a number of towns,[85] but in Exeter, too, it was the prospect of a British Association meeting which gave supporters of the measure the leverage they needed to win over the doubters. At a public meeting in the town held just five months before the BAAS was due to arrive, mayor Henry Ellis and his supporters managed to carry the motion to introduce the Act by a four to one majority.[86]

Developments in library provision and in the facilities and membership of provincial scientific societies helped build the foundations for more ambitious education schemes in some provincial towns in the last quarter of the nineteenth century. The 1870s and 1880s saw the establishment of a number of new colleges delivering university education in towns such as Newcastle, Leeds, Bristol, Sheffield, Birmingham, Liverpool, Nottingham and Dundee. Leeds made a bid to establish itself as a county centre for science education with the establishment of the Yorkshire College of Science, thanks to money from the Cloth Workers' Company of London and other donors.[87] Similarly, Bristol's new college, established in 1876, was styled the College of Science for the West of England, although its first principal had ambitions to 'give the business man of Bristol that true literary education which refines the mind and broadens its interests in human

---

83 *History, Gazetteer and Directory of Norfolk, 1883* (Sheffield, 4th edition, 1883), p. 486.

84 *Birmingham Daily Post*, 2 September 1865.

85 See for example, Geoffrey Drewery, *History and Development of Libraries in Hull* (Hull: Kingston Press, 2002), pp. 32–7.

86 For a full account of the meeting see Westcountry Studies Library [WSL], WEB LIB150, Exeter Free Library: public meeting, 6 May 1869.

87 Arnold N. Shimmin, *The University of Leeds. The First Half Century* (Cambridge, 1954), p. 13.

life'.[88] Sheffield's Firth College and Birmingham's Mason's College, established in 1879 and 1880 respectively, both owed their origins to a generous benefactor keen to promote education in science and technology.[89] The University College Liverpool, which opened a year later after a series of public meetings and funding appeals which yielded over £100,000, was broader based in its curriculum, with a commitment from the outset to teach non-scientific subjects.[90]

Historians of the English university sector have acknowledged the debt of the civic university movement to the cultural institutions of towns in the mid-Victorian era. Research on the origins of a number of university colleges has identified the role of mechanics' institutes, medical schools and libraries in fostering interest and in providing practical facilities such as accommodation and resources to support university level education in provincial towns.[91] As Alberti has pointed out in his work on university colleges in Victorian Britain, there was a crucial link with early nineteenth-century provincial literary and scientific institutions and their members, many of whom were instrumental in providing the practical support and lobbying necessary.[92] In Newcastle for example, the Literary and Philosophical Society helped provide lecture rooms for the college of medicine from 1834.[93] In Nottingham, the town's Mechanics Institute, established in 1837, was 'the embryo from which, forty years on, the University College was to spring'.[94] Consideration of the connections between the development of the civic universities in the last quarter of the nineteenth century and the earlier scientific and literary life of provincial towns has not previously encompassed the staging of national scientific congresses. Yet as part of the knowledge fabric of urban Britain in the period, the parliaments of science played an integral part in creating the conditions and opportunities to foster the development of university education.

There is evidence to suggest a strong connection between towns which were successful in hosting parliaments of science and those which were among the first to develop university colleges. Six of the eight English towns in which university institutions were established between 1851 and 1882 were among the

---

88 A. Marshall, 'Some Aspects of Modern Industrial Life', quoted in Michael Sanderson (ed.), *The Universities in the Nineteenth Century* (London, 1975), pp. 161–2.

89 Jones, *Origins of Civic Universities*, p. 18.

90 Jones, *Origins of Civic Universities*, pp. 58–62.

91 See for example, Jones, *The Origins of Civic Universities*, pp. 19–21; Samuel J.M.M. Alberti, 'Civic Cultures and Civic Colleges in Victorian England', in M. Daunton (ed.), *The Organisation of Knowledge in Victorian Britain* (Oxford, 2005), pp. 334–56.

92 Samuel J.M.M. Alberti, 'Amateurs and Professionals in One County: Biology and Natural History in Late Victorian Yorkshire', *Journal of the History of Biology*, 34 (2001): 125–7.

93 Ernest M. Bettenson, *The University of Newcastle-upon-Tyne, 1834–1971. A Historical Introduction* (Newcastle, 1971), p. 14.

94 Alfred C. Wood, *A History of the University College of Nottingham, 1881–1948* (Oxford, 1953), p. 5.

top 20 locations for annual meetings featured in Table 2.2 in Chapter 2,[95] and a closer examination of the paths taken by some of these places towards university establishments suggests that this was no coincidence. In many of them, a degree of overlap in personnel is evident between those who played a role in attracting scientific meetings to these towns and the central figures in later educational developments. In Newcastle upon Tyne, for example, where a medical school had been in existence since 1834, a science college was established in 1871 partly on the initiative of two leading figures in the town's BAAS meetings, local industrialists William Armstrong and Isaac Lowthian Bell, with the assistance of Dean Lake of Durham University and Michael Wood of the Institute of Mining and Mechanical Engineers. Its original professors were appointed in Mathematics, Chemistry, Physics and Geology and the institution was conceived as a science college which would serve the northern counties of England.[96] Elsewhere, too, there were direct connections between the personnel involved in annual meetings and those who went on to campaign for university institutions in their towns. Contemporaries saw annual meetings as events which stimulated interest in university education. After the British Association's meeting in Dundee in 1867, William Knight, Professor of Moral Philosophy of St Andrew's from 1876, reflected that, 'The meeting … in Dundee deeply stirred the community and guided the desire for higher education. At many a gathering of the citizens and many a social meeting, the question was discussed'.[97] It was significant that among the local organising committee for the Dundee BAAS meeting in 1867 was John Boyd Baxter, who later donated £10,000 towards the cost of establishing the university college, which opened in 1883.[98] In Nottingham there was also an overlap in personnel between the executive committee responsible for organising the BAAS visit in 1866 and the source of benefactions which helped establish the university college some 15 years later. Lace curtain manufacturer Lewis Heymann and his son Albert were both members of the organising committee for the BAAS meeting in 1866. A third member of the same family, William Heymann, was believed to be the source of an anonymous donation of £10,000, which allowed educational developments in the town to be set on a more ambitious footing.[99] The corporation had already acquired library and museum collections previously housed in voluntary institutions in the town, and university extension lectures were instituted in the town in 1875,[100] but with

95 The two others, Nottingham and Sheffield, each hosted British Association annual meetings, in 1866 and 1879 respectively.

96 Bettenson, pp. 21–2.

97 Quoted in, A.D. Walsh, 'Notes on the first visit of the British Association to Dundee and of the Early History of University College Dundee', in S.J. Jones (ed.), *Dundee and District* (Dundee, 1968), p. 374.

98 Walsh, 'Notes on the first visit', p. 374–8.

99 Church, *Economic and Social Change in a Midland Town*, p. 363.

100 John V. Beckett and B.H. Tolley, 'Church, Chapel and School', in John V. Beckett (ed.), *A Centenary History of Nottingham* (Chichester, 2006), p. 377.

the Heymann gift, it was able to press ahead with arrangements for a building in which its library and museum collections could be housed and in which university lectures could take place by 1881.[101]

As the Nottingham case shows, the establishment of a civic university required an accumulation of effort, resource and enthusiasm over time towards the goal of higher educational facilities. The experience of hosting a major knowledge congress could play an important part in this process, and not just because of the platform it provided for local scientists and other educational enthusiasts. Also important was the taste it gave the town of the benefits of collaborating with an influential outside body to advance its educational ambitions. In some places this experience usefully foreshadowed the crucial final stepping stone on the road to university establishment: the adoption of the University Extension Scheme. The Scheme was initiated by James Stuart, a fellow of Trinity College, Cambridge, who was motivated by 'the widespread and real desire for some form of higher education which existed throughout the country' and his belief in 'the obligation that there was on the two ancient universities to come forward to supply that demand'.[102] Under the scheme, Oxford and Cambridge lecturers taught classes in provincial centres. It meant that towns could offer a much wider range of instruction than would otherwise have been possible. In Leeds, for example, university extension lectures enabled a broadening of the curriculum of its college, sufficient for affiliation to the newly established Victoria University in 1887.[103] In Sheffield, the success of university extension lectures went hand in hand with the founding of Firth College which, as one contemporary acknowledged, 'in reality is the child of the Extension Scheme'.[104]

Stuart's scheme was, from its inception, 'a joint movement of the universities and towns' in which the 'university undertakes the educational organisation, the town the funds and local management: the whole constituting a network of local branches working independently, in association with the universities as a common centre'.[105] In other words, it required of participating towns an organisational relationship with a central body which was very similar to the kind of partnership struck up each year between host venue and national association in preparation for an annual meeting, and it comes as no surprise to find some of the same people involved in the operation of both. In Leeds, for example, the local Extension Committee was chaired by Edward Baines, MP and newspaper proprietor, who had a successful track record of bringing knowledge organisations to the town through his participation in delegations from Leeds to the British Association and the Royal Agricultural Society. His efforts had helped to bring both organisations

[101] Wood, *A History of the University College of Nottingham*, pp. 8–23.

[102] J. Stuart, 'Reminiscences', quoted in Sanderson, *The Universities in the Nineteenth Century*, pp. 180–182.

[103] Shimmin, *The University of Leeds*, pp. 16–17.

[104] Anonymous letter to the editor, *York Herald*, 29 July 1879.

[105] Richard G. Moulton, *The University Extension Movement* (London, 1885), p. 5.

to the town in 1858 and 1861 respectively. It is easy to see why the University Extension movement flourished in some of the places which had experience of hosting national learned congresses and it is possible, from this perspective, to view the parliaments of science template, which saw liaison between a town-based local organising committee and representatives of a central association as a kind of blueprint for the University Extension Scheme.

Yet in addition to these common organisational threads, the experience of hosting learned meetings was more broadly influential on the movement for civic colleges because it was part of a repertoire of educational and cultural activity in provincial towns which helped associate cultural developments with the bolstering of civic pride and urban status. Henry Clifton Sorby FRS, talented natural historian and stalwart of the Sheffield Literary and Philosophical Society, articulated as much in 1878 when he wrote to the *Sheffield and Rotherham Independent* on the subject of 'Sheffield's Reputation'. His fellow townsmen, he asserted, would 'be gratified to know in what high terms Sheffield is now spoken of in London':

> What we have lately done in connection with the examinations and scholarships to encourage education in the elementary schools, the remarkable success of the University lectures and classes, and of the Gilchrist lectures, the Firth College and the manner in which the British Association appears likely to be received in Sheffield have already produced a most satisfactory effect … I was extremely gratified to hear at the late reception at the Royal Society a well-known professor of Cambridge speak of Sheffield as being 'the leading city in the country' in such matters as those I have alluded to … Such a recognition, amongst many others, of what we have already done will, I hope, encourage us in our efforts to promote the intellectual as well as the material advancement of our town.[106]

The initiatives he referred to had transformed the educational facilities of the town which one educational guidebook, published just a decade earlier, had characterised as lagging behind others, with 'fewer literary and educational institutions than they might support and their demands upon the bookseller and the librarian are not in proportion to their numbers or their need of knowledge'.[107] By 1879, however, other Yorkshire towns were looking on with envy at what Sheffield had achieved. One correspondent to the *York Herald* argued that York's pedigree, as the first meeting place of the British Association and the home of the Yorkshire Philosophical Society, ought to provide the foundations for similar achievements.[108]

---

[106] *Sheffield and Rotherham Independent*, 12 March 1878.

[107] *Sheffield. Published under the direction of the committee of general literature and education appointed by the society for promoting Christian Knowledge* (London, 1864), p. 30.

[108] Anonymous letter to the editor, *York Herald*, 29 July 1879.

Such sentiments reveal that British Association meetings were seen by at least some contemporaries as part of an interwoven fabric of cultural activity which gave valuable momentum to the movement towards university education in the provinces. The case of Exeter provides a further illustration of how this worked in practice. When the town played host to the 1869 British Association meeting, the experience provided a tangible boost to cultural life in Exeter. As we have seen, the prospect of the arrival of BAAS members helped local promoters to gain support for the adoption of the Free Library Act and the completion of the Albert Memorial Museum, while afterwards, buoyed by success of the meeting, this latter institution, which housed the Schools of Art and Science, was taken over by the city. In 1886, when the Albert Memorial committee decided to apply to Cambridge for a course of university extension lectures, the ambition of the town to improve the quality and reach of its adult education classes was confirmed and the pace of change quickened with the establishment of the Exeter Technical and University Extension College in 1893.[109] In 1900 it was renamed the Royal Albert Memorial College, but became known to the general public as the University College Exeter and eventually gained university college status in 1922 as the University College of the South West of England.[110]

Looked at in this context, parliaments of science can be understood as events which helped set some towns on the path towards establishing themselves as regional centres for education and learning. Awareness that the fostering of knowledge and cultural life could enhance town status long predated the visitation of national associations in some towns. In Nottingham, for instance, the sense of civic pride in such activities was evident well before the visit of the British Association in 1866. One mid-century guidebook writer sang the praises of the town's progress over earlier decades, claiming that through the establishment of 'noble literary, artistic, philosophical and benevolent institutions – Nottingham has emphatically merged into a Midland Metropolis, in mind and manufactures, in a mental as well as a material aspect'.[111] As elsewhere, however, the British Association visit helped bring these assets into sharper relief. With reputations riding on the outcome of these meetings and visiting delegates ready to compare facilities and arrangements with those experienced in previous host towns, the minds of municipal leaders were concentrated on the condition of local scientific societies, museum and library accommodation, and the hosting of congresses were, as a number of examples in this chapter have shown, the motivating cause behind the upgrading of many such facilities in British towns. In turn, the kudos and experience gained by scientific men and urban leaders in the locality of a meeting can be traced in the later activities of some of the same people in pressing for higher education in the provincial towns. The hosting of parliaments of science was only one strand in fabric of cultural

[109] Brian W. Clapp, *The University of Exeter: A History* (Exeter, 1982), p. 20.

[110] Clapp, pp. 57–63.

[111] William H. Wylie, *The Nottingham Hand-book and Guide to Places of Interest in the Environs* (Nottingham, 1852), p. iii.

activity in which the towns were engaged in the mid-Victorian period, but it provided a connecting thread which helped concentrate efforts and resources on the promotion of learning and instilled urban leaders with a confidence and belief to pursue knowledge provision as a key urban function.

Conclusion

In 1884 the British Association took its annual meeting outside the British Isles for the first time when it selected the Canadian city of Montreal as its venue.[1] This move met with considerable opposition from sections of the British membership of the BAAS, but the support of the Governor General of Canada and the precedent of other meetings, including that of the American Association for the Advancement of Science, which met in Montreal in 1848, was sufficient to win the day.[2] In the opinion of Philip Lowe, however, it was an indication that the BAAS by this date was struggling to find new stimulus and direction.[3] With British towns now well equipped with museums, libraries and, in some cases, with universities, it was more difficult for the Association to think of itself as an enlightening influence, even in towns at a considerable distance from the metropolis. Consequently there was, according to the editors of *The Times*, a sense of stagnation about the Association's activities:

> It cannot be denied that the British Association is to some extent on its trial at the present time. It is passing through an ordeal such as comes as a matter of course to every established institution at a certain stage of its existence. It has long outlived the attraction of mere novelty, and is compelled in its turn to pay the penalty which attaches to popularity.[4]

The imperial dimension opened up new horizons and brought a new sense of mission, to foster closer links with overseas dominions. When the Association returned to Canada for its second visit in Victoria's Diamond Jubilee year, *The Times* proclaimed that the 'conception of the unity of the Queen's dominions has become more deeply rooted than ever' and urged the Association to re-brand itself 'the scientific parliament of the British Empire'.[5]

The first overseas meeting of the BAAS seemed to mark a natural break in the cycle of annual parliaments of science for the previous 50 years and provides a useful point from which to reflect on the phenomenon in Britain's provincial towns. The British Association's experience corresponded with shifting patterns of organisation in the other associations too. In the same year that BAAS members

---

1 Editorial comment on the Montreal meeting can be found in *The Times*, 16 August 1884.

2 Howarth, pp. 119–21.

3 Lowe, 'The British Association and the Provincial Public', p. 132.

4 *The Times*, 25 September 1883.

5 *The Times*, 5 August 1897.

were travelling to Montreal, the NAPSS held its last annual congress, back where it began, in Birmingham.[6] The loss of key individuals had taken its toll on the Social Science Association. Lord Brougham died in 1868 and Hastings was elected to parliament in 1880, thus diverting his considerable energies away from the Association and its work.[7] Falling attendances had been in evidence since the late 1870s. The Manchester meeting of 1879, for example, was blighted by bad weather which 'sadly militated against the popular success of the Congress'. By the mid-point of the week 'there had been issued at the local offices 598 tickets … a small total compared with many previous gatherings of the association' and it was noted that in some of the sectional meetings 'there were very few people present'.[8] It was a pale imitation of the previous Social Science meeting in the town 13 years earlier which had attracted an attendance of over 1,800. Even more than was the case with the BAAS, the NAPSS found its role as a promoter of social reform altered as the ground shifted around it. The 1880s was the beginning of a period of significant extension of local government activity as municipalities took over responsibility for a wide range of services and facilities in the towns, from the management of water supplies, transport and libraries to sanitary inspections and, later, slum clearance and the provision of new housing.[9] The corresponding rise in the number of local government employees with specialist expertise in these spheres, increasingly with their own dedicated representative associations, diminished the need for an umbrella organisation to share knowledge and lobby for legislation.

The other national bodies were also operating on shifting sands by the 1880s. RASE had a range of changing social and economic circumstances to contend with. The agricultural depression of the period cast a shadow over its council meetings and journal publications, but membership and attendance at the annual shows remained buoyant. Like the other associations featured in this study it was feeling the effects of the passage of time. A number of key long-time members died, including Augustus Voelcker in 1884 and Brandreth Gibbs in 1885. A gradual ebbing away of the optimism which had accompanied the foundation of RASE was also palpable. By the 1880s, divisions between town and country were growing and the interest and engagement of urban dwellers with agricultural matters was

---

6 It should be noted that the Association sponsored a conference on temperance legislation in London in 1886 and consequently its cessation is sometimes dated from this point. See for example Ronald K. Huch, 'The National Association for the Promotion of Social Science: its contribution to Victorian health reform, 1857–1886', *Albion*, 17, 3 (1985), p. 294.

7 Further discussion on this can be found in Goldman, *Science, Reform and Politics*, pp. 349–67.

8 MCL, BR F306 F2, *Manchester Examiner*, 4 October 1879.

9 Barry M. Doyle, 'The changing functions of urban government: councillors, officials and pressure groups, in Martin Daunton (ed.), *CUHB volume III, 1840–1950* (Cambridge, 2000), pp. 287–96.

on the wane.[10] In 1879 a financially disastrous annual show at Kilburn, blighted by heavy rain throughout the week, coupled with the depressed economic state of agriculture, led to a period of reflection and soul-searching as to the future of the Society. *The Agricultural Gazette* concluded that:

> It has had its share hitherto in the work of the century, and has carried out its first programme in uniting science with practice, and stimulating invention in every department of farming. We do not by any means suppose that its mission has been fulfilled and its work finished, but we do say that the position of farming has entirely changed since the Society commenced its labours ... Their work consists no longer in the invention of machinery, and breeds of animals, and chemical manures, and in the careful comparison of novelty in show yards. Their duty lies – especially in the case of proprietors – at home in the attraction to the land of the necessary capital, and in rendering the tenantry secure.[11]

At the same time, there were growing levels of state involvement in agricultural affairs. News that a minister for agriculture was to be appointed by the government was greeted with scepticism in some sections of the farming community who feared that the effect would be to 'revive the hopes of the far too large a body in our midst who think that a great deal can be done for us by the Government which we cannot do for ourselves'.[12]

The Royal Archaeological Institute continued to operate well into the twentieth century but it too found its *modus operandi* affected by social and institutional changes in the 1870s and 1880s. The death of long-time president, Talbot de Malahide, in 1883 seemed to symbolise the end of its first generation and the Institute subsequently entered a difficult period characterised by falling membership.[13] All around there were new influences and authorities with pressing interests in archaeology. The Society for the Protection of Ancient Buildings, founded in 1877 by William Morris, attracted an influential membership advocating the conservation, rather than the restoration, of ancient buildings.[14] The passage of the Ancient Monuments Protection Act of 1882 which appointed an inspector of ancient monuments and enabled the state to take on guardianship of historic sites in need of protection[15] set the tone for the next decade as newly re-organised local government bodies began to take on more responsibility for archaeological preservation. Two other trends had a profound influence on the context in which

---

10 See Goddard, *Harvests of Change*, pp. 116–38.

11 *The Agricultural Gazette*, 14 July 1879.

12 *The Agricultural Gazette*, 14 July 1879.

13 Ebbatson, thesis, p. 68.

14 Charles Dellheim, *The Face of the Past. The Preservation of the Medieval Inheritance in Victorian England* (Cambridge, 1982), pp. 84–92.

15 Timothy Champion, 'Protecting the monuments: archaeology legislation from the 1882 Act to PPG16', in Hunter (ed.), *Preserving the Past*, pp. 38–41.

the RAI was operating by the last quarter of the nineteenth century. One was the flourishing state of local and county archaeological societies and the work of a great number of them in investigating and preserving important sites. According to the historian of the RAI, this level of activity in the localities came at the expense of the development of a national framework and standard for the discipline and had the effect of prolonging the amateur status of archaeology.[16] It was an issue which archaeologists attempted to address through a process of re-organisation and the convening of the Congress of Archaeological Societies in 1888 which aimed to centralise arrangements for the collection and storage of artefacts.[17] The other trend was the growing academic status of historical knowledge. The formation of the Royal Historical Society in 1868 marked the separation of historical research from wider antiquarian interests and a widening gulf between the professional status of the former and the amateur connotations of the latter. The foundation of the *English Historical Review* in 1886, meanwhile, signalled the 'academic arrival of the professional historian'.[18] The careers of individuals like Edward Freeman demonstrated how these factors played out in practice. An active RAI member who had participated in numerous summer meetings as a speaker and a guide, he bristled with indignation when one provincial newspaper referred to his work in 'compiling' a history of the Norman Conquest,[19] presumably because of the implication that it involved little more than an exercise in data collecting. He ended his career in a university post, as Regius Professor of History at Oxford in 1884.[20]

The new influences making their effects felt on the Archaeological Institute were also evident in other spheres of knowledge. The increasing professionalisation and specialisation of intellectual activity and the development of distinct academic disciplines was evident not just in Britain but elsewhere in both Europe and America. One consequence of this was that the community became less of a focus for the communication of new ideas, as leading practitioners turned instead to their peer group through the pages of academic journals, specialist conferences and academic university departments. This shift from amateur, community-based knowledge to a more professional, disciplinary approach has been described as 'the great story of nineteenth-century science',[21] but it was not just a feature of scientific knowledge. Professionalisation was a general societal trend from the 1880s, evident in the increasing specialisation of work, in a proliferation of professional associations and in a huge growth in government.[22] The implications

16 Ebbatson, thesis, p. 15.

17 Levine, *The Amateur and the Professional*, p. 53; Ebbatson, thesis, p. 242.

18 Levine, *The Amateur and the Professional*, p. 164.

19 Edward W. Freeman to Dean Hook, 3 September 1867, quoted in Stephens (ed.), *Life and Letters*, vol. 1, p. 391.

20 Ebbatson, thesis, p. 60.

21 Bender, *Intellect and Public Life*, pp. 20–21.

22 Harold Perkin, *The Rise of Professional Society. England since 1880* (London, 1989). pp. 20–22.

for national knowledge associations and their annual meetings were far reaching. No longer could they claim to be the main repositories of learning and ideas. In fact, in a society which increasingly valued specialised expertise, associations designed to embrace multiple branches of knowledge, and to facilitate contact between specialists and 'amateurs', risked appearing out of step. Moreover, communication with one's academic peers could now be done without the need for a mass annual gathering. The occasion of the 50th anniversary of the British Association for the Advancement of Science in 1881 led the editors of *The Times* to reflect that the personal communication afforded through the meeting,

> was of very different importance in 1831 from what is is now, when the post goes everywhere in a night and Edinburgh is but ten hours from London. The yearly gatherings … were expected and remembered as great opportunities for seeing friends from a distance and talking over difficulties with them … Nowadays a chemist or spectroscopist in Edinburgh or Manchester can in a few hours be in the library of the Royal Society or British Museum …[23]

Given these conditions, it is possible to argue that the heyday of annual parliaments of science was over by the mid-1880s. This is especially so if we see them as a product of the strong Liberal consensus of the mid-Victorian period.[24] Quite apart from the shared belief in the towns in the reforming and enlightening mission of the associations, the organisational feat of hosting them relied, as we saw in Chapter 3, on the coming together of representatives from different groups and their willingness to unite around a common scientific aim. It was difficult for institutions built on this bedrock to survive the 'earthquake of the 1880s' when the Liberal Party split and the interests of land and business capital began to ally themselves more firmly with the Conservatives.[25] At local government level, all of this was happening against a backdrop of re-structuring in the late 1880s. The Local Government Act of 1888 extended the principle of elected councils to the counties by establishing county councils made up of councillors, elected for three years, and aldermen, in office for six years.[26] The effect, according to one contemporary commentator, was to complete 'the work begun in 1835 in the reform of municipal corporations. Now rural as well as urban life rests on a representative basis'.[27] The growing sense of accountability to ratepayer electors

---

23 *The Times*, 26 August 1881.

24 This argument is made in the case of the National Association for the Promotion of Social Science, Lawrence Goldman, 'The Social Science Association, 1857–1886: a context for mid-Victorian Liberalism', *English Historical Review*, 101/398 (1986): 95–134.

25 Perkin, p. 40.

26 J.A. Chandler, *Explaining Local Government. Local Government in Britain Since 1800* (Manchester, 2007), p. 100.

27 Dr Pankhurst, 'The Provinces under the Local Government Act', in *Pall Mall Gazette*, 8 September 1888.

and an unwillingness to commit town funds to causes which served the interests of private or privileged groups can be seen permeating discussions on the merits of hosting society meetings by this period. In Nottingham in 1887, for example, two council members opposed the signing of a guarantee to find the requisite subscription required to host the Royal Agricultural Society. They argued that, 'seeing that the landowners and agriculturalists would reap most benefit from the show, that the Mayor ought not to be authorised to sign a document which would throw responsibility for a large proportion of the expenditure upon the poor ratepayers of Nottingham'.[28] It was a comment that expressed much more than just local sensitivities; it gave voice to the increasing sense of separate interest groups in society with unequal shares of wealth and influence.

In the light of these sentiments in the towns, the cost of hosting meetings was becoming an increasingly contentious issue by the last quarter of the nineteenth century. As we saw in Chapter 2, the preparedness of so many towns to compete for the right to host an annual meeting, and to give over their buildings, hospitality and resources to the cause, was what gave these events their prestige and their sustainability in the mid-Victorian years. Without this level of interest on the part of the towns, the future of peripatetic meetings could quickly become uncertain. RASE began to recognise, by the 1870s, that the escalating size and cost of its annual shows was becoming a deterrent to some towns to consider hosting. In 1874, RASE's finance committee estimated the total cost of an annual show at over £15,000.[29] At the time the Society was considering venues for its 1875 show in the south west, but only Taunton came forward as a prospective host. By the end of the nineteenth century RASE was considering the merits of abandoning the peripatetic method in favour of a permanent showground. A fixed site at Twyford, known as Park Royal, was tried for three years from 1903 to 1905 and, although unsuccessful, the experiment confirmed that rotating provincial shows were not regarded as having an indefinite shelf-life.[30] Other associations also experienced a waning of enthusiasm on the part of the towns. In the difficult economic and industrial conditions of the 1870s and 1880s, the drive of one or two eager individuals was no longer always enough to persuade cost-conscious councillors that they should foot the bill for such events. Portsmouth town council withdrew its invitation to host the annual social science congress in 1885 because of concerns over the size of the guarantee fund required by the Association. The town council had been advised that a sum in the region of £820 would be necessary and, although not as high as the costs incurred at some congresses, this was thought to be too much. Following a special meeting of the town council, a resolution was

---

[28] *Nottinghamshire Guardian*, 11 March 1887.

[29] MERL, SR RASE BII/4 Minutes of Finance Committee, 1870–1887, 4 November 1874.

[30] Full details of the Park Royal episode can be found in Goddard, *Harvests of Change*, pp. 139–48.

drawn up stating its regret that, 'the amount required by the Committee of the Social Science Association ... cannot be obtained for 1885'.[31]

On the face of it this was a clear case of belt-tightening, suggesting that towns, for the sake of economy, were willing to forego their week in the limelight entertaining the members of knowledge organisations which were, in any case, too big and too broad in their remit for more specialised, contemporary tastes. But this does not tell the whole story. From the perspective of the provincial towns there were other issues to consider, as the case of Portsmouth reveals. The withdrawal of the invitation to the Social Science Association by Portsmouth's mayor came about not just because of the likely costs of the event, but because the town had what it saw as a better offer on the table. The Church Congress, also interested in holding an annual meeting in Portsmouth in 1885, had the potential to draw in a crowd of around 2,500 to the town, but was likely to cost only £200 or £300 because the organisation covered more of its own expenses. Councillors spoke up enthusiastically about the prospect of hosting this event because 'churches and chapels were springing up all around them, and as there was much religious enthusiasm exhibited the whole of the money required would no doubt be soon raised'.[32] This was just one of a number of tempting prospects on the horizon. In August 1885, a public meeting was held to discuss the propriety of inviting the Royal Counties' Agricultural Society (of Hampshire and Berkshire) to Portsmouth for its 1886 meeting, an event which required a guarantee fund of £450.[33] The county agricultural societies were part of the growing plethora of regional associations looking for places to host their meetings in the last quarter of the nineteenth century. County archaeological societies and 'regional replicas' of the British Association for the Advancement of Science, such as those in Devon and Yorkshire had, to some extent, become an alternative focus for the ambitions of towns to stage annual events. The lower costs involved did not mean a reduced potential for attracting large visitor numbers from surrounding districts, and hence the county events were an attractive prospect.

It was not just Portsmouth where townsmen found themselves weighing up the pros and cons of staging a national parliament of science against the merits of alternative events. In 1884, Leicester town council was approached by a group of architects in the town with a request to invite the Sanitary Institute to hold its 1885 conference in Leicester. Councillors noted that the Institute was concerned with 'the advancement of sanitary science and the diffusion of knowledge relating thereto; and it was especially interesting to the public because its members were not merely professional men, but included men like the Earl of Shaftesbury, Sir John Lubbock, and others whose interests were chiefly scientific and philanthropic'.[34] Having satisfied themselves that the event would not be too specialised to ignite

---

31 *Hampshire Telegraph and Sussex Chronicle*, 13 December 1884.

32 Ibid.

33 *Hampshire Telegraph and Sussex Chronicle*, 29 August 1885.

34 *Leicester Chronicle and Leicestershire Mercury*, 30 August 1884.

public interest, and with the added reassurance that the required guarantee fund would amount to no more than £500, councillors decided to proceed with issuing an invitation. This was despite the fact that they were already, 'to some extent engaged towards making an invitation to another association, very much larger and very much older, and one which had a much wider reputation altogether – the British Association'.[35] Having already received one refusal from the BAAS which favoured Aberdeen for its 1885 meeting, and with no guarantee of a favourable response, the Leicester men showed little hesitation in looking elsewhere for opportunities to play host.

As these cases demonstrate, Victorian towns had not lost their appetite for showcasing knowledge by the last quarter of the nineteenth century, but they were using a greater variety of means to do it. As well as the opportunity to hold county events and host organisations like the Church Congress and the Sanitary Institute, much bigger opportunities loomed into view for towns with an appetite for spectacle, which dwarfed even the largest of the mid-Victorian parliaments of science. These were the international exhibitions of industries and manufactures on the 1851 Crystal Palace model. From 1855 to 1914, grand exhibitions featuring the industries of multiple nations were staged on average once every two years somewhere in the world.[36] In Britain, the original plan of using the South Kensington facilities to stage further great exhibitions on a decennial basis was superseded by the 1870s with instigation of yearly exhibitions of a more specialised nature.[37] Anticipating the forthcoming international fisheries exhibition of 1883, *The Times* noted that:

> An exhibition is now considered, and probably rightly considered, the most effectual mode of pushing an industry or promoting certain kinds of reform. Whether the object be to raise up Irish or Yorkshire manufactures, or to induce people to consume their own smoke, an exhibition is the magician's wand that is to work the wonder.[38]

Similar international exhibitions of health and of forestry, as well as those promoting the industries of particular regions, took place throughout the decade, rendering the international exhibition a staple of annual calendar by the 1880s. Keen to emulate the success of the original Crystal Palace event, major European and American cities hosted their own versions, often timed to coincide with an

---

35 Ibid.

36 Paul Greenhalgh, *Ephemeral Vistas: The Expositions Universelles, Great Exhibitions and World's Fairs, 1851–1939* (Manchester, 1988), p. 15. The difficulty of applying a precise definition to these events, however, has led to different estimates by other scholars. See for example the 35 international expositions identified in Gold and Gold (eds), *Cities of Culture*, pp. 77–8.

37 For details see letters from J. Forbes Watson to *The Times*, 28 December and 30 December 1872.

38 *The Times*, 27 February 1882.

anniversary or other 'celebratory circumstance', in order to achieve maximum impact.[39] Exhibition mania also spread to a number of provincial British cities where the resources were found to assemble bewildering arrays of exhibits in lavish buildings at immense cost. The spectacular Glasgow International Exhibition at the city's Kelvingrove Park in 1888, for example, covered an area of 474,500 square feet and contained 22 different classes of exhibit, ranging from naval architecture, to road carriages and cooking implements, and from jewellery and textiles to fine arts.[40] Such events did not come cheap. The guarantee fund for the event totalled £300,000: raised from contributions from 'the commercial men of the West of Scotland' and by throwing open the event to participants and contributors from much further afield.[41] Small wonder that these events, with their great visual attractions and their wide public appeal, went some way toward eclipsing scientific meetings in their ability to excite interest and occupy urban resources and facilities by the 1880s.

International exhibitions on the scale of the one staged in Glasgow in 1888 were years in the planning as well as being enormously costly. They represented one extreme on an increasingly diverse scale of place promotion initiatives open to towns by the last quarter of the nineteenth century. Another prize which became available to a much wider range of towns in the late 1880s, and one which could be attained through little more than the preparing of a petition and the payment of a fee, was that of city status. Manchester had been granted the title of city in 1853 after a campaign initiated during Queen Victoria's visit two years earlier, but few towns had tried to follow suit in the 1850s and 1860s. At the time, cities were places designated as the see of an Anglican bishop; there was no great sense of what general advantage such a designation might convey. In any case, towns were occupied with acquiring other signifiers of status in the form of their public buildings and institutions. By the latter years of the century, however, 'city status had been transformed from a distinction attached to a town with a cathedral, to a mark of civic esteem'.[42] The removal of the association with Anglican bishoprics came in 1888 and, subsequently, there was a clearer attempt to define cities as places of elevated importance, based partly on population size. The message soon filtered through to provincial towns that city status was a distinction within their grasp. Birmingham town councillors, noting that 'The title of city is popularly supposed to carry with it a certain accession of dignity, and is no longer held to be necessarily connected with the see of a bishop', used the occasion of the 50th anniversary of their municipal charter to make a case for city status. They based their claim on Birmingham's position as, 'the second largest town in England (excluding London)' and its record of municipal government over 50 years which, it claimed, 'can hardly have been surpassed'.[43] It was this growing

---

39 Greenhalgh, *Ephemeral Vistas*, p. 16.

40 *The Dundee Courier and Argus*, 6 April 1888.

41 *Aberdeen Weekly Journal*, 20 September 1887.

42 John V. Beckett, *City Status in the British Isles, 1830–2002* (Aldershot, 2005), p. 19.

43 *Birmingham Daily Post*, 30 November 1888.

understanding of cities as places of greater importance than towns which meant that city status 'finally came by the 1890s to be seen as a necessary accessory for the great provincial towns of industrial Britain'.[44] By the outbreak of war in 1914, the towns of Dundee, Birmingham, Leeds, Sheffield, Bradford, Nottingham, Hull and Cardiff, none of them the location of Anglican sees, had joined Liverpool, Newcastle-upon-Tyne and Manchester in the ranks of major industrial provincial centres now designated cities. It was confirmation, if any was still needed, that these populous, commercial and industrial centres had risen through the urban ranks to positions that distinguished them from the traditional market and county centres.

The range of alternative avenues through which urban status could be acquired and promoted did not bring an end to the staging of annual meetings for the national associations in provincial towns. Indeed, by the 1880s, new influences were brought to bear on the way in which these events were staged in some locations. In Birmingham, for example, the British Association visit of 1886 was organised alongside the Birmingham Exhibition of Industrial Products and Processes which, the mayor claimed, 'would see a number of processes carried on by machinery in motion which, he believed, had but only once been equalled under any roof, and that was in the Exhibition of 1851'.[45] This inter-weaving of different strands of place promotion was one means by which towns helped adapt the annual meeting formula to meet the demands of changing times. An investigation of the full extent of this process of adaptation lies outside the scope of the present volume. Rather, it is the contention of this book that the years from the 1830s to the 1880s, when the parliaments of science occupied a prominent and distinctive place in the civilising agenda of towns, merit attention in their own right. In this 50-year period there was a convergence of a number of key influences which made this possible. The ambition and status-consciousness of provincial towns; the commitment of learned associations to knowledge dissemination through itinerant annual meetings; the desire of this top tier of learned associations to be seen as 'national' in outlook and nation-enhancing in their aims; and the development of facilities for the transportation and accommodation of large numbers of people around the country, all combined to make the annual scientific meeting an especially potent tool, both for furthering scientific and civic ambitions and thereby enlightening the nation.

Yet it is a half century which has too often occupied a gap in the historiography, between the study of the earlier 'lit. and phil'. movement in the provinces,[46] and the later era of the rise of civic universities.[47] In fact, the parliament of science years from the 1830s to the 1880s had strong linkages both forward and

---

[44] Beckett, *City Status*, p. 41.

[45] *Birmingham Daily Post*, 27 August, 1886.

[46] See for example, Elliott, 'Towards a geography of English scientific culture'; Inkster and Morrell (eds), *Metropolis and Province*; G.S. Kitteringham, 'Science in provincial society: the case of Liverpool in the early nineteenth century', *Annals of Science*, 39/4 (1982): 329–48.

[47] Jones, *The Origins of the Civic Universities.*

backward, building on earlier urban scientific traditions, and paving the way for later educational developments. One objective of this study has been to establish this interconnectivity between these different phases of knowledge organisation and to go some way towards filling the gap in the picture of scientific activity in nineteenth-century towns. Fundamentally though, it is the towns themselves which lie at the heart of this book. Filled with people who viewed the acquisition of scientific knowledge as a route to betterment of self and society, Victorian towns were already doing, in miniature, what the national knowledge associations of the day wanted to achieve country-wide. By partnering with these bodies the towns' efforts were magnified to national level. Suddenly their distinguished scientists, their museum spaces, the collections assembled by their townsmen, their local feats of engineering, manufacture and their archaeological and architectural treasures became objects of nationwide scrutiny and admiration. The power of this was not lost on provincial urbanites who were looking for means to assert themselves as, all around, neighbouring places vied for urban supremacy. But far from passively receiving the national associations and their members, the towns made these events their own, embellishing them with layers of civic ceremony, local expertise and municipal hospitality. If Britain's parliaments of science were invented by the BAAS and its successor organisations, they were adopted and shaped by the towns of the provinces. They came to form an integral part of the repertoire of cultural initiatives and events from which towns sought to bolster their status and shape their urban identity. More than just the places of meeting, host towns became active agents in the shaping of the major scientific congresses of the Victorian era. The town-focused perspective offered in this book is not intended as a replacement for history of science interpretations of the activities of the BAAS and its fellow knowledge organisations in nineteenth century Britain, but rather as an assertion that these events have a place in the realms of urban history too.

Appendicies

Appendix 1 Alphabetical list of towns hosting scientific meetings, 1831–1884

Host Town	BAAS	RAI	RASE	NAPSS	TOTAL
Aberdeen	1	0	0	1	2
Bath	1	1	0	0	2
Bedford	0	1	1	0	2
Belfast	2	0	0	1	3
Birmingham	3	0	0	3	6
Bradford	1	0	0	1	2
Brighton	1	0	0	1	2
Bristol	2	1	1	1	5
Bury St Ed-munds	0	1	1	0	2
Cambridge	3	1	1	0	5
Canterbury	0	2	1	0	3
Cardiff	0	1	1	0	2
Carlisle	0	2	1	0	3
Chelmsford	0	0	1	0	1
Cheltenham	1	0	0	1	2
Chester	0	1	1	0	2
Chichester	0	1	0	0	1
Colchester	0	1	0	0	1
Cork	1	0	0	0	1
Derby	0	0	1	0	1
Dorchester	0	1	0	0	1
Dublin	3	0	0	2	5
Dundee	1	0	0	0	1
Edinburgh	3	1	0	2	6
Exeter	1	1	1	0	3
Glasgow	3	0	0	2	5
Gloucester	0	1	1	0	2
Hereford	0	1	0	0	1
Huddersfield	0	0	0	1	1
Hull	1	1	1	0	3
Ipswich	1	0	0	0	1
Lancaster	0	1	0	0	1

Appendix 1 *Continued*

Host Town	BAAS	RAI	RASE	NAPSS	TOTAL
Leeds	1	0	1	1	3
Leicester	0	1	1	0	2
Lewes	0	1	1	0	2
Lincoln	0	1	1	0	2
Liverpool	3	0	1	2	6
London*	0	2	2	1	5
Manchester	2	0	1	2	5
Newcastle	2	2	2	1	7
Northampton	0	1	1	0	2
Norwich	1	1	1	1	4
Nottingham	1	0	0	1	2
Oxford	3	1	2	0	6
Peterborough	0	1	0	0	1
Plymouth	2	0	1	1	4
Reading	0	0	1	0	1
Ripon	0	1	0	0	1
Rochester	0	1	0	0	1
Salisbury	0	1	1	0	2
Sheffield	1	0	0	1	2
Shrewsbury	0	1	2	0	3
Southampton	2	1	1	0	4
Southport	1	0	0	0	1
Swansea	2	0	0	0	2
Taunton	0	1	1	0	2
Warwick	0	1	1	0	2
Winchester	0	1	0	0	1
Windsor	0	0	1	0	1
Wolverhamp-ton	0	0	1	0	1
Worcester	0	1	1	0	2
York	3	1	1	1	6

* Note that RASE shows in Battersea (1862) and Kilburn (1879) appear here as 'London'

Appendix 2 Locations of annual scientific meetings by year, 1831–1884

Year	BAAS	NAPSS	RAI	RASE
1831	York			
1832	Oxford			
1833	Cambridge			
1834	Edinburgh			
1835	Dublin			
1836	Bristol			
1837	Liverpool			
1838	Newcastle			
1839	Birmingham			Oxford
1840	Glasgow			Cambridge
1841	Plymouth			Liverpool
1842	Manchester			Bristol
1843	Cork			Derby
1844	York		Canterbury	Southampton
1845	Cambridge		Winchester	Shrewsbury
1846	Southampton		York	Newcastle
1847	Oxford		Norwich	Northampton
1848	Swansea		Lincoln	York
1849	Birmingham		Salisbury	Norwich
1850	Edinburgh		Oxford	Exeter
1851	Ipswich		Bristol	Windsor
1852	Belfast		Newcastle	Lewes
1853	Hull		Chichester	Gloucester
1854	Liverpool		Cambridge	Lincoln
1855	Glasgow		Shrewsbury	Carlisle
1856	Cheltenham		Edinburgh	Chelmsford
1857	Dublin	Birmingham	Chester	Salisbury
1858	Leeds	Liverpool	Bath	Chester
1859	Aberdeen	Bradford	Carlisle	Warwick
1860	Oxford	Glasgow	Gloucester	Canterbury
1861	Manchester	Dublin	Peterborough	Leeds
1862	Cambridge	London	Worcester	Battersea
1863	Newcastle	Edinburgh	Rochester	Worcester
1864	Bath	York	Warwick	Newcastle
1865	Birmingham	Sheffield	Dorchester	Plymouth
1866	Nottingham	Manchester	London	
1867	Dundee	Belfast	Kingston-upon-Hull	Bury St Edmunds
1868	Norwich	Birmingham	Lancaster	Leicester
1869	Exeter	Bristol	Bury St Edmunds	Manchester

Appendix 2 ***Continued***

Year	BAAS	NAPSS	RAI	RASE
1870	Liverpool	Newcastle	Leicester	Oxford
1871	Edinburgh	Leeds	Cardiff	Wolverhampton
1872	Brighton	Plymouth	Southampton	Cardiff
1873	Bradford	Norwich	Exeter	Hull
1874	Belfast	Glasgow	Ripon	Bedford
1875	Bristol	Brighton	Canterbury	Taunton
1876	Glasgow	Liverpool	Colchester	Birmingham
1877	Plymouth	Aberdeen	Hereford	Liverpool
1878	Dublin	Cheltenham	Northampton	Bristol
1879	Sheffield	Manchester	Taunton	Kilburn
1880	Swansea	Edinburgh	London	Carlisle
1881	York	Dublin	Bedford	Derby
1882	Southampton	Nottingham	Carlisle	Reading
1883	Southport	Huddersfield	Lewes	York
1884	Montreal	Birmingham	Newcastle	Shrewsbury

Bibliography

Manuscript Sources

Bodleian Library

Dep. BAAS 1 Correspondence of John Phillips.
Dep. BAAS 17 Council Minutes, 1832–1842.
Dep. BAAS 59 Correspondence of John Phillips.
Dep. BAAS 130–33 Notebooks of John Phillips.
Dep. BAAS 142 General Correspondence relating to annual meetings.
Dep. BAAS 146–55 Printed material relating to annual meetings.

Bristol Record Office

29427 (1–6) Booklets and printed materials relating to Bristol meetings of the BAAS in 1898 and 1930.
32079(39) Minute Book of the Reception Committee for the British Association visit.
37454 (15) d Autograph book.
38605/X/5 Mark Whitwill's Report of the Social Science Congress held at Glasgow, 30 Sept. 1874.

Manchester Central Library

BR F306 N1 National Association for the Promotion of Social Science, circulars etc.
BR F306 F2 Social Science Association, Manchester Meeting 1879.
BR F306 S3 Social Science Association, Manchester Meeting 1866.

Museum of English Rural Life

SR RASE B/XXXII/1 Letterbook, 1838–1844.
SR RASE B/I/1 Minutes of Council 1840–1842 (rough copy).
SR RASE B/I/2 Minutes of Council 1842–1844 (rough copy).
SR RASE B/I/3 Minutes of Council 1844–1848 (rough copy).
SR RASE B/I/4 Minutes of Council 1848–1853 (rough copy).
SR RASE B/I/5 Minutes of Council 1853–1858 (rough copy).
SR RASE B/I/6 Minutes of Council 1858–1868 (rough copy).

SR RASE BI/7 Minutes of Council 1869–1881 (rough copy).
SR RASE B1/10 Minutes of the Committee of Management (fair copy).
SR RASE BII/4 Minutes of Finance Committee, 1870–1887.
SR RASE B/III/2 Agenda of Country Committees, 1851–1864.
SR RASE, B/III/ 3 Agenda of Country Committees, 1864–1876.
SR RASE B/III/ 4 Agenda of Country Committees, 1876–1887.
SR RASE B/XIII/4 Miscellaneous.

National Museums Liverpool, Maritime Archives and Library

Danson Family Archive: John Towne Danson Collection.
D/D/III/2/62 Letters from George W. Hastings to John Towne Danson.
D/D/III/8/3/3 Report of the Liverpool Committee of the NAPSS, 1859.
D/D/III/15/3 National Association for the Promotion of Social Science.
D/D/III/17/5a Newspaper Cutting Book, 1858–1861.

Society of Antiquaries Library

WAY SAL/MS/700/VII.
WAY MS/790.
SAL/MS/877/1.

Tyne and Wear Archives and Museums

TWAM DF COW/D/241 Joseph Cowan notebook.
TWAM MD NC/94/3 General Committee Book no.3A.

University College London, Special Collections

Brougham MSS.

Westcountry Studies Library

WES/1869 D'Urban BAAS, Exeter Meeting, August 1869. Map edited by W.S.M. D'Urban, FLS, for the Local Committee.
WEB LIB150 Exeter Free Library: public meeting, 6 May 1869.

Online Resources

Carlyle Letters Online http://carlyleletters.org.

Printed Primary Sources

'Annual Meeting for 1871 to be Held at Cardiff', *Archaeological Journal*, 27 (1870).

Armstrong, William G., Isaac Lowthian Bell, J. Taylor and Dr Richardson (eds), *The Industrial Resources of the District of the Three Northern Rivers, the Tyne, Wear and Tees* (London: Longman, 2nd edition, 1864).

Beavan, Alfred B., *Bristol Lists: Municipal and Miscellaneous* (Bristol: T.D. Taylor Sons and Hawkins, 1899).

Clark, John W., *The Life and Letters of the Rev. Adam Sedgwick*, vols I and II (Cambridge: Cambridge University Press, 1890).

Eighteenth Meeting of the British Association for the Advancement of Science. President's address, 9 August 1848.

Felkin, William, *A History of the Machine-Wrought Hosiery and Lace Manufactures* (London: Longmans, Green and Co., 1867),

Gamwell, Samuel C., *The Official Guide and Handbook to Swansea and its District* (Swansea: British Association for the Advancement of Science, 1880).

Geikie, Archibald, *Life of Sir Roderick Murchison, Based on his Journals and Letters* (London: Murray, 1875).

History, Gazetteer and Directory of Norfolk, 1883 (Sheffield: William White, 4th edition, 1883).

PP 1863 LIII pt.1, *Census of England and Wales, 1861. General Report.*

Pengelly, Hester (ed.), *A Memoir of William Pengelly of Torquay, FRS, Geologist, with a selection from his correspondence* (London: John Murray, 1897).

Planché, James Robinson, *Recollections and Reflections* (new and revised edition, London: Sampson Low, Marston and Company Ltd, 1901).

Post Office Bradford Directory, 1879–80 (Bradford: W. Byles and Son, 1879).

Proceedings and Reports of the Town Council of the Borough of Newcastle-upon-Tyne (Newcastle: M. and M.W. Lambert, various years).

Records of the Borough of Nottingham. Being a Series of Extracts from the Archives of the Corporation of Nottingham, vol. IX: 1836–1900 (Nottingham: Thomas Forman and Sons Ltd, 1956).

Report of the Second Meeting of the Devonshire Association for the Advancement of Science Literature and Art, held at Plymouth, July 1863 (London, 1864).

Report of the Twelfth Meeting of the British Association for the Advancement of Science held at Manchester in June 1842 (London, 1843).

Report of the Eighteenth Meeting of the British Association for the Advancement of Science (London, 1849).

Report of the Twenty-Third Meeting of the British Association for the Advancement of Science held at Hull in September 1853 (London, 1854).

Report of the Thirty-Third Meeting of the British Association for the Advancement of Science, held at Newcastle Upon Tyne In August and September 1863 (London, 1864).

Report of the Thirty-Fifth Meeting of the British Association for the Advancement of Science, held at Birmingham in September 1865 (London, 1866).

Report of the Thirty-Eighth Meeting of the British Association for the Advancement of Science, held at Norwich in August 1868 (London, 1869).

Russell, Bertrand and Patricia (eds), *The Amberley Papers. Bertrand Russell's Family Background*, vol. 1 (London: Allen and Unwin, 1937).

Sheffield. Published under the Direction of the Committee of General Literature and Education Appointed by the Society for Promoting Christian Knowledge (London: S.P.C.K., 1864).

Shepherdson, William, *The New Hull Guide: A Modern Picture of Hull and Visitor's Handbook to the Various Objects of Interest in Hull and Neighbourhood* (Hull: M.C. Peck and Son, 1875).

Social Science: Being Selections from John Cassell's Prize Essays by Working Men and Women with Notes (London, 1859).

Stephens, William R.W. (ed.), *The Life and Letters of Edward A. Freeman*, vol. 1 (London: Macmillan and Co., 1895).

Symons, John, *The Visitors' Descriptive Guide Book to Kingston upon Hull* (Hull: William Adams; London: Kent and Co., 1867).

Timmins, Samuel (ed.), *The Resources, Products and Industrial history of Birmingham and the Midlands Hardware District* (London: R. Hardwicke, 1866).

Trades' Societies and Strikes. Report of the Committee on Trades' Societies Appointed by the National Association for the Promotion of Social Science (London: John W. Parker and Son, 1860).

Transactions of the National Association for the Promotion of Social Science, Sheffield Meeting (London: Longman, Roberts and Green, 1866).

Wylie, William H., *The Nottingham Hand-book and Guide to Places of Interest in the Environs* (Nottingham: Christopher N. Wright jun., 1852).

Newspapers

Aberdeen Journal
Agricultural Gazette
Belfast Newsletter
Berrow's Worcester Journal
Birmingham Daily Post
Bradford Observer
Bristol Mercury
Bury and Norwich Post and Suffolk Herald
Caledonian Mercury
Cambrian
Cardiff and Merthyr Guardian
Chester Observer and General Advertiser

Daily News
Derby Mercury
Devon Weekly Times
Dundee Courier and Argus
Exeter Flying Post
Glasgow Herald
Hampshire and Salisbury Guardian
Hull Packet
Illustrated London News
Illustrated Times
Ipswich Journal
Jackson's Oxford Journal
Lancaster Gazette and General Advertiser
Leeds Mercury
Liverpool Mercury
Lloyd's Weekly Newspaper
Manchester Times and Gazette
Morning Chronicle
Morning Post
Newcastle Courant
Nottinghamshire Guardian
Pall Mall Gazette
Penny Illustrated Paper and Illustrated Times
Preston Guardian
Sheffield and Rotherham Independent
The Era
The Graphic
The Standard
The Times
Western Mail
York Herald and General Advertiser

Secondary Sources

Alberti, Samuel J.M.M., 'Amateurs and Professionals in One County: Biology and Natural History in Late Victorian Yorkshire', *Journal of the History of Biology*, 34 (2001): 115–47.

Alberti, Samuel J.M.M., 'Conversaziones and the Experience of Science in Victorian England', *Journal of Victorian Culture*, 8/2 (2003): 208–30.

Allen, David Elliston (ed.), *Naturalists and Society. The Culture of Natural History in Britain, 1700–1900* (Aldershot: Ashgate, 2001).

Allison, Keith J. (ed.), *A History of the County of York East Riding: Volume 1: The City of Kingston upon Hull* (London: Institute of Historical Research, 1969).

Ayris, Ian, *A City of Palaces. Richard Grainger and the Making of Newcastle-upon-Tyne* (Newcastle: Newcastle Libraries and Information Services, 1997).

Beckett, John V., *City Status in the British Isles, 1830–2002* (Aldershot: Ashgate, 2005).

Beckett, John V. (ed.), *A Centenary History of Nottingham* (Chichester: Phillimore, 2006).

Berry, Helen, 'Creating Polite Space: The Organisation and Social Function of the Newcastle Assembly Rooms', in Berry, Helen and Gregory, Jeremy, (eds), *Creating and Consuming Culture in North-East England, 1660–1830* (Aldershot: Ashgate, 2004).

Bender, Thomas, *New York Intellect. A History of Intellectual Life in New York City From 1750 to the Beginning of Our Own Times* (New York: Alfred A. Knopf, 1987).

Bender, Thomas, *Intellect and Public Life. Essays on the Social History of Academic Intellectuals in the US* (Baltimore: John Hopkins University Press, 1993).

Best, Geoffrey, *Mid-Victorian Britain, 1851–75* (London: Fontana, 1971).

Bettenson, Ernest M., *The University of Newcastle-upon-Tyne, 1834–1971. A Historical Introduction* (Newcastle: University of Newcastle upon Tyne, 1971).

Bingham, Adrian, 'The Digitization of Newspaper Archives: Opportunities and Challenges for Historians', *Twentieth Century British History*, 21/2 (2010): 225–31.

Birch, Neville, 'The "Archaeologians" Visit Lincoln', in Sturman, Christopher, (ed.), *Lincolnshire People and Places: essays in memory of Terence R. Leach (1937–1994)* (Lincoln: Society for Lincolnshire History and Archaeology, 1996).

Borsay, Peter N., 'The English Urban Renaissance: The Development of Provincial Urban Culture, 1680–1760', *Social History*, 5 (1977): 581–603.

Brake, Laurel and Demoor, Marysa, (eds), *The Lure of Illustration in the Nineteenth Century* (Basingstoke: Palgrave, 2009).

Briggs, Asa, *The Age of Improvement 1783–1867* (Harlow: Longman, 7th edition, 1988).

Bunting, Brian T., 'Morton, John Chalmers (1821–1881)', *Oxford Dictionary of National Biography* (Oxford: Oxford University Press, 2004).

Burn, William Laurence, 'Newcastle-upon-Tyne in the Early Nineteenth Century', *Archaeologia Aeliana*, 4th series, 34 (1956): 1–13.

Burn, William Laurence, *The Age of Equipoise. A Study of the Mid-Victorian Generation* (London: Allen and Unwin, 2nd edition, 1968).

Burnley, James, 'Fairbairn, Sir William, First Baronet (1789–1874)', Rev. R. Brown, *Oxford Dictionary of National Biography* (Oxford: Oxford University Press, 2008).

Campbell, W.A., 'Men of Science in Nineteenth-century Newcastle', *Archaeologia Aeliana*, 5th series, 8 (1980): 57–64.

Cannadine, David, 'The Transformation of Civic Ritual in Modern Britain: The Colchester Oyster Feast', *Past and Present*, 94 (February, 1982): 107–30.

Cappock, Margarita, 'Pageantry or Propaganda? The Illustrated London News and Royal Visitors in Ireland', *Irish Arts Review Yearbook*, 16 (2000): 86–93.

Carter, Charles Frederick, (eds), *Manchester and its Region* (Manchester: Manchester University Press, 1962).

Chalklin, Christopher W. and Wordie, J. Ross (eds), *Town and Countryside. The English Landowners in the National Economy* (London, 1989).

Chandler, J.A., *Explaining Local Government. Local Government in Britain since 1800* (Manchester: Manchester University Press, 2007).

Chapman, Stanley D., 'Felkin, William (1795–1874)', *Oxford Dictionary of National Biography* (Oxford: Oxford University Press, 2004).

Charleton, Robert J., *A History of Newcastle-on-Tyne. From the Earliest Records to its Formation as a City* (Newcastle: William H. Robinson, 1885).

Church, Roy A., *Economic and Social Change in a Midland Town: Victorian Nottingham, 1815–1900* (London: Frank Cass, 1966).

Clapp, Brian W., *The University of Exeter: A History* (Exeter: University of Exeter, 1982).

Clark, Peter (ed.), *The Transformation of English Provincial Towns, 1500–1800* (London: Hutchinson, 2nd edition, 1985).

Clark, Peter (ed.), *The Cambridge Urban History of Britain volume II, 1540–1840* (Cambridge, Cambridge University Press, 2000).

Clark, Peter, *British Clubs and Societies, 1580–1800. The Origins of an Associational World* (Oxford: Oxford University Press, 2001).

Clarke, Ernest, 'Shaw, William (1797–1853)', Rev. Goddard, Nicholas P.W., *Oxford Dictionary of National Biography* (Oxford: Oxford University Press, 2004).

Clifford, David, Warwick, Alex, Wadge, Elisabeth, and Willis, Martin (eds), *Repositioning Victorian Sciences. Shifting Centres in Nineteenth-century Scientific Thinking* (London: Anthem Press, 2006).

Cooper, M.L. and Hall, V.D.L., 'William Robert Grove and the London Institution', *Annals of Science*, 39 (1982): 229–54.

Cooper, Victoria L., 'Novello, Clara Anastasia (1818–1908)', *Oxford Dictionary of National Biography* (Oxford, 2004).

Corfield, Penelope, 'Georgian Bath – the Magical Meeting Place', *History Today*, 40/11 (1990): 26–33.

Corfield, Penelope, 'From Second City to Regional Capital', in Rawcliffe, Carole and Wilson, Richard G., (eds), *Norwich Since 1550* (London: Hambledon, 2004).

Cottrell, A.E., *The History of Clifton Suspension Bridge* (Bristol: A.E. Cottrell, 6th edn., 195).

Crimmin, P.K., 'Duncan, Adam, Viscount Duncan (1731–1804), *Oxford Dictionary of National Biography* (Oxford: Oxford University Press, 2004).

Crosby, Alan G., *The Chester Archaeological Society. The First One Hundred and Fifty Years, 1849–1999* (Chester: Chester Archaeological Society, 1999).

Cunningham, Colin, *Victorian and Edwardian Town Halls* (London: Routledge and Kegan Paul, 1981).

Daunton, Martin (ed.), *Cambridge Urban History of Britain volume III, 1840–1950* (Cambridge: Cambridge University Press, 2000).

Daunton, Martin (ed.), *The Organisation of Knowledge in Victorian Britain* (Oxford: Oxford University Press, 2005).

De Nie, Michael W., *The Eternal Paddy: Irish Identity and the British Press, 1798–1882* (Madison: University of Wisconsin Press, 2004).

Dellheim, Charles, *The Face of the Past. The Preservation of the Medieval Inheritance in Victorian England* (Cambridge: Cambridge University Press, 1982).

Dierig, Sven, Lachmund, Jens, and Mendelsohn, J. Andrew (eds), *Science and the City. Osiris*, 18 (2003).

Elliott, Paul, 'Towards a Geography of English Scientific Culture: Provincial Identity and Literary and Philosophical Culture in the English County Town, 1750–1850', *Urban History*, 32/3 (2005): 391–412.

Ellis, Joyce, 'For the Honour of the Town: Comparison, Competition and Civic Identity in Eighteenth-century England, *Urban History*, 30/3 (2003): 325–37.

Emerson, Roger L., 'The Enlightenment and Social Structures', in Fritz, Paul and Williams, David (eds), *City and Society in the Eighteenth Century* (Toronto: Hakkert, 1973).

Evans, Joan, 'The Royal Archaeological Institute: A Retrospect', *Archaeological Journal*, 106 (1951 for 1949): 1–11.

Everitt, Alan, 'Country, County and Town: Patterns of Regional Evolution in England', in Borsay, Peter N. (ed.), *The Eighteenth-century Town. A Reader in English Urban History, 1688–1820* (London, 1990).

Fairbrother, R., Birks, J.B., Wolfe Mays, and Morgan, P.G., 'The History of Science in Manchester', in C.F. Carter (ed.), *Manchester and its Region* (Manchester, 1962), pp. 187–97.

Fawcett, Trevor, 'Popular Science in Eighteenth-century Norwich', *History Today*, 22 (1972), pp. 590–95.

Feinstein, Charles, (ed.), *York 1831–1981. 150 Years of Scientific Endeavour and Social Change* (York: Sessions, 1981).

Finnegan, Diarmid A., 'Natural History Societies in late Victorian Scotland and the Pursuit of Civic Science', *British Journal for the History of Science*, 38/1 (2005): 53–72.

Finnegan, Diarmid A., 'The Spatial Turn: Geographical Approaches in the History of Science', *Journal of the History of Biology*, 41 (2008): 369–88.

Gill, Conrad, *History of Birmingham, Volume I: Manor and Borough to 1865* (London: Oxford University Press, 1952).

Gill, Crispin, *Plymouth. A New History* (Tiverton: Devon Books, 1993).

Goddard, Nicholas P.W., *Harvests of Change. The Royal Agricultural Society of England, 1838–1988* (London: Quiller, 1988).

Gold, John R. and Gold, Margaret M., *Cities of Culture. Staging International Festivals and the Urban Agenda, 1851–2000* (Aldershot: Ashgate, 2005).

Gold, John R. and Gold, Margaret M., 'Culture and the City', *History in Focus*, 13 (2008).

Gold, John R. and Ward, Stephen V. (eds), *Place Promotion. The Use of Publicity and Marketing to Sell Towns and Regions* (Chichester: Wiley, 1994).

Goldman, Lawrence, 'The Social Science Association, 1857–1886: A Context for Mid-Victorian Liberalism', *English Historical Review*, 101/398 (1986): 56–134.

Goldman, Lawrence, *Science, Reform and Politics in Victorian Britain. The Social Science Association, 1857–1886* (Cambridge: Cambridge University Press, 2002).

Golinski, Jan V., 'Utility and Audience in Eighteenth-century Chemistry: Case Studies of William Cullen and Joseph Priestly', *The British Journal for the History of Science*, 21/1 (1988): 1–31.

Greenaway, Frank, 'Dalton, John (1766–1844)', *Oxford Dictionary of National Biography* (Oxford University Press: Oxford, 2004; online ed., Oct. 2006).

Greenaway, Frank, 'Henry, William (1774–1836), *Oxford Dictionary of National Biography* (Oxford University Press: Oxford, 2004, online ed., 2007).

Greenhalgh, Paul, *Ephemeral Vistas: The Expositions Universelles, Great Exhibitions and World's Fairs, 1851–1939* (Manchester: Manchester University Press, 1988).

Gunn, Simon, 'Ritual and Civic Culture in the English Industrial City, c.1835–1914', in Morris, Robert J. and Trainor, Rick H. (eds), *Urban Governance. Britain and Beyond Since 1750* (Aldershot: Ashgate, 2000).

Gunn, Simon, *The Public Culture of the Victorian Middle Class. Ritual and Authority in the English Industrial City, 1840–1914* (Manchester: Manchester University Press).

Halcrow, E.M., 'The Town Moor of Newcastle upon Tyne', *Archaeologia Aeliana*, 4th series, 31 (1953): 149–64.

Harrison, Carol E. and Johnson, Ann, 'Introduction: Science and National Identity', *Osiris*, 24/1, *Science and National Identity* (2009): 1–14.

Harrison, Mark, *Crowds and History. Mass Phenomena in English Towns, 1790–1835* (Cambridge: Cambridge University Press, 1988).

Hart, Gwen, *A History of Cheltenham* (Leicester: Leicester University Press, 1965).

Hewitt, Martin (ed.), *An Age of Equipoise? Re-assessing Mid-Victorian Britain* (Aldershot: Ashgate, 2000).

Higgitt, Rebekah and Withers, Charles W.J., 'Science and Sociability. Women and Audience at the British Association for the Advancement of Science, 1831–1901', *Isis*, 99/1 (2008): 1–27.

Hill, Francis, *Victorian Lincoln* (Cambridge: Cambridge University Press, 1974).

Howard, Jill, '"Physics and Fashion": John Tyndall and his Audiences in Mid-Victorian Britain', *Studies in the History and Philosophy of Science Part A*, 35/4 (2004): 729–58.

Hoock, Holger, 'From Beefsteak to Turtle: Artists' Dinner Culture in Eighteenth-Century London', *Huntington Library Quarterly*, 66/1–2 (2003): 27–54.

Howarth, Osbert J.R., *The British Association for the Advancement of Science: A Retrospect, 1831–1931* (London, 1931).

Huch, Ronald K., 'The National Association for the Promotion of Social Science: its Contribution to Victorian Health Reform, 1857–1886', *Albion*, 17/3 (1985): 298.

Hunter, Michael (ed.), *Preserving the Past. The Rise of Heritage in Modern Britain* (Stroud: Alan Sutton, 1996).

Inkster, Ian and Morrell, Jack (eds), *Metropolis and Province. Science in British Culture, 1780–1850* (London, 1983).

Inkster, Ian, *Scientific Culture and Urbanisation in Industrial Britain* (Aldershot: Ashgate, 1997).

James, Frank A.J.L., 'An "Open Clash between Science and the Church"? Wilberforce, Huxley and Hooker on Darwin at the British Association, Oxford, 1860', in David M. Knight and Matthew D. Eddy (eds), *Science and Beliefs. From Natural Philosophy to Natural Science, 1700–1900* (Aldershot: Ashgate, 2005).

James, K.A., '"Humbly Dedicated": Petiver and the Audience for Natural History in Early Eighteenth-century Britain', *Archives of Natural History*, 31/2 (2004): 318–29.

Jobey, George, 'The Society of Antiquaries of Newcastle-upon-Tyne', *Archaeologia Aeliana*, 5th series, 18 (1990): 197–216.

John, Angela V. 'Crawshay [née Yeates], Rose Mary (1828–1907)', *Oxford Dictionary of National Biography* (Oxford: Oxford University Press, 2004).

Jones, David R., *The Origins of the Civic Universities. Manchester, Leeds and Liverpool* (London: Routledge, 1988).

Jones, Howard Mumford and Cohen, I. Bernard (eds), *Science Before Darwin. A Nineteenth-Century Anthology* (London: Deutsch, 1963).

Jordanova, Ludmilla, 'Science and Nationhood: Cultures of Imagined Communities', in Geoffrey Cubitt (ed.), *Imagining Nations* (Manchester: Manchester University Press, 1998).

Kargon, Robert H., *Science in Victorian Manchester. Enterprise and Expertise* (Manchester: Manchester University Press, 1977).

Kellett, John R., *The Impact of Railways on Victorian Cities* (London: Routledge and Kegan Paul, 1969).

Kidd, Alan and Nicholls, David (eds), *Gender, Civic Culture and Consumerism. Middle-class Identity in Britain, 1800–1940* (Manchester: Manchester University Press, 1999).

Kitteringham, G.S., 'Science in Provincial Society: The Case of Liverpool in the Early Nineteenth Century', *Annals of Science*, 39/4 (1982): 329–48.

Kingston, Ralph, 'Mind Over Matter? History and the Spatial Turn', *Cultural and Social History*, 17/1 (2010): 111–21.

Kitson Clark, George Sidney Roberts, *The Making of Victorian England* (London: Methuen, 1962).

Levine, Philippa, *The Amateur and the Professional. Antiquarians, Historians and Archaeologists in Victorian England, 1833–1886* (Cambridge: Cambridge University Press, 1986).

Linsley, Stafford M., 'Armstrong, William George, Baron Armstrong (1810–1900)', *Oxford Dictionary of National Biography* (Oxford: Oxford University Press, 2006).

Little, Ernest M., *History of the British Medical Association, 1832–1932* (London: British Medical Association, 1932).

Livingstone, David N., *Putting Science in its place. Geographies of Scientific Knowledge* (Chicago: University of Chicago Press, 2003).

Lloyd-Morgan, Glenys, 'The Early Growth of Archaeology in Chester: 1849–1895', *Journal of the Chester Archaeological Society*, 72 (for 1992/3): 15–23.

Looney, J.J., 'Cultural Life in the Provinces. Leeds and York, 1720–1820', in Beier, A.L., Cannadine, David, and Rosenheim, James M. (eds), *The First Modern Society. Essays in English History in Honour of Lawrence Stone* (Cambridge: Cambridge University Press, 1989).

Lowe, Philip, 'The British Association and the Provincial Public', in MacLeod, Roy and Collins, Peter (eds), *The Parliament of Science. The British Association for the Advancement of Science, 1831–1981* (Northwood, 1981).

McGregor, O.R., 'Social Research and Social Policy in the Nineteenth Century', *British Journal of Sociology*, 8/2 (1957): 146–57.

Macilwee, Michael, *The Liverpool Underworld: Crime in the City, 1750–1900* (Liverpool: Liverpool University Press, 2011).

Mandler, Peter, '"The Wand of Fancy": The Historical Imagination of the Victorian Tourist', in Marius Kwint, Christopher Breward, and Aynsley, Jeremy (eds), *Material Memories. Design and Evocation* (Oxford: Berg, 1999).

Miskell, Louise, 'The Making of a New "Welsh Metropolis": Science, Leisure and Industry in Early Nineteenth-century Swansea', *History*, 88/1 (2003): 32–52.

Miskell, Louise, *Intelligent Town. An Urban History of Swansea* (Cardiff: University of Wales Press, 2006).

Miskell, Louise, 'Meeting Places: The Scientific Congress and the Host Town in the South West of England, 1836–1877', *Urban History*, 39/2 (2012): 246–62.

Mitchell, Rosemary, 'Cassell, John (1817–1865)', *Oxford Dictionary of National Biography* (Oxford: Oxford University Press, 2004).

Moore, James R., 'Urban Space and Civic Identity in Manchester, 1780–1914: Piccadilly Square and the Art Gallery Question', *Transactions of the Historic Society of Lancashire and Cheshire*, 153 (2004): 87–123.

Morley, Ian, 'Representing a City and a Nation: Wales' Matchless Civic Centre', *Welsh History Review*, 24/3 (2009): 56–81.

Morrell, Jack, 'The Early Yorkshire Geological and Polytechnic Society: a reconsideration', *Annals of Science*, 45/2 (1988): 153–67.

Morrell, Jack, 'Wissenschaft in Worstedopolis: Public Science in Bradford, 1800–1850', *British Journal of the History of Science*, 18 (1985): 1–23.

Morrell, Jack, 'Harcourt, William Venables Vernon (1789–1871)', *Oxford Dictionary of National Biography* (Oxford: Oxford University Press, 2004).

Morrell, Jack and Thackray, Arnold, *Gentlemen of Science. Early Years of the British Association for the Advancement of Science* (Oxford: Clarendon Press, 1981).

Morrell, Jack and Thackray, Arnold, *Gentlemen of Science. Early Correspondence of the British Association for the Advancement of Science* (London: Royal Historical Society, 1984).

Morus, Iwan Rhys, 'Manufacturing Nature: Science, Technology and Victorian Consumer Culture', *British Journal for the History of Science*, 29/4 (1996): 403–34.

Morus, Iwan Rhys, 'Grove, Sir William Robert (1811–1896)', *Oxford Dictionary of National Biography* (Oxford: Oxford University Press, 2005).

Morus, Iwan Rhys, 'Worlds of Wonder: Sensation and the Victorian Scientific Perfomance', *Isis*, 101/4 (2010): 806–16.

Moulton, Richard G., *The University Extension Movement* (London: Benrose and Sons, 1885).

Murphy, Sherra, 'The Learned Gentlemen are in Town: The British Association Meeting of 1857 in Dublin's Popular Press', in Adelman, Juliana and Agnew, Éadaoin, (eds), *Science and Technology in Nineteenth-century Ireland* (Dublin: Four Courts, 2011).

Musgrove, Frank, 'Patterns of Intellectual Creativity in the North of England', *Northern History*, 33 (1997): 1–35.

Naylor, Simon, 'Nationalizing Provincial Weather: Meteorology in Nineteenth-century Cornwall', *British Journal for the History of Science*, 39/3 (2006): 407–33.

Naylor, Simon, *Regionalizing Science: Placing Knowledges in Victorian England* (London: Pickering and Chatto, 2010).

Neal, Frank, *Sectarian Violence. The Liverpool Experience, 1819–1914. An Aspect of Anglo-Irish History* (Manchester: Manchester University Press, 1988).

Newey, Vincent, 'Robinson, Henry Crabb (1775-1867)', *Oxford Dictionary of National Biography* (Oxford, 2007).

Nurse, Bernard, 'Way, Albert (1805–1874)', *Oxford Dictionary of National Biography* (Oxford: Oxford University Press, 2004).

Nuttall, Robert H., 'The Achromatic Microscope in the History of Nineteenth-century Science', *The Philosophical Journal*, 11 (1974): 71–88.

O'Gorman, Frank, 'Campaign Rituals and Ceremonies: The Social Meaning of Elections in England, 1780–1860', *Past and Present*, 135 (1992): 79–115.

Oddy, Derek J., 'Urban Famine in Nineteenth-century Britain. The Effect of the Lancashire Cotton Famine on Working-class Diet and Health', *Economic History Review*, new series, 36/1 (1983): 68–86.

Ofcansky, Thomas Paul, 'Baker, Sir Samuel White (1821–1893)', *Oxford Dictionary of National Biography* (Oxford: Oxford University Press, 2004; online ed. 2008).

Ogawa, Mariko, 'Leibig and the Royal Agricultural Society of England meeting at Bristol, 1842', *Ambix*, 85/2 (2008): 136–52.

Orange, A.D., 'The Idols of Theatre: The British Association and its Early Critics', *Annals of Science*, 32/3 (1975): 278–94.

Otter, Chris, *The Victorian Eye. A Political History of Light and Vision in Britain, 1800–1910* (Chicago: Chicago University Press, 2008).

Owen, John A., *The History of the Dowlais Ironworks, 1759–1970* (Merthyr Tydfil: Merthyr Tydfil Borough Corporation, 1973).

Paradis, James G., 'Satire and Science in Victorian Culture', in Lightman, Bernard (ed.), *Victorian Science in Context* (Chicago: Chicago University Press, 1997).

Parissien, S.P. 'Tite, Sir William (1798–1873)', *Oxford Dictionary of National Biography* (Oxford: Oxford University Press, 2004).

Peacock, A.E., 'The Creation of the West Riding Court of Assize', *Northern History*, 22 (1987): 119–37.

Perkin, Harold, *The Rise of Professional Society. England Since 1880* (London: Routledge, 1989).

Poole, Steve, 'To be a Bristolian: Civic Identity and the Social Order, 1750–1850', in Dresser, Madge and Ollerenshaw, Philip (eds), *The Making of Modern Bristol* (Tiverton: Redcliffe, 1996).

Prochaska, Frank, 'Kinnaird, Arthur Fitzgerald, Tenth Lord Kinnaird of Inchture and Second Baron Kinnaird of Rossie (1814–1887)', *Oxford Dictionary of National Biography* (Oxford: Oxford University Press, 2004).

Robson, Ann P., 'Russell, Katharine Louisa, Viscountess Amberley (1842–1874)', *Oxford Dictionary of National Biography* (Oxford: Oxford University Press, 2004).

Rodger, Richard, 'The "Common Good" and Civic Promotion: Edinburgh 1860–1914', in Robert Colls and Richard Rodger (ed.), *Cities of Ideas. Civil Society and Urban Governance in Britain, 1800–2000. Essays in Honour of David Reeder* (Aldershot: Ashgate, 2004).

Rothery, Mark, 'The Shooting Party: The Associational Cultures of Rural and Urban Elites in the Late Nineteenth and Early Twentieth Centuries', in Richard W. Hoyle (ed.), *Our Hunting Fathers. Field Sports in England after 1850* (Lancaster: Carnegie, 2007).

Russell, Colin A., *Science and Social Change, 1700–1900* (London: Macmillan, 1983).

Sanderson, Michael (ed.), *The Universities in the Nineteenth Century* (London: Routledge and Kegan Paul, 1975).

Shapin, Steven, 'The Audience for Science in Eighteenth-century Edinburgh, *History of Science*, 12/2 (1974): 95–121.

Shimmin, Arnold N., *The University of Leeds. The First Half Century* (Cambridge: Cambridge University Press, 1954).

Shteir, Ann B., *Cultivating Women, Cultivating Science: Flora's Daughters and Botany in England, 1760-1860* (Baltimore: John Hopkins University Press, 1996).

Smith, Crosbie, 'Joule, James Prescott (1818–1889)', *Oxford Dictionary of National Biography* (Oxford: Oxford University Press, 2011).

Spurrell, Mark (ed.), *Stow Church Restored, 1846–1866. Lincolnshire Record Society Publications*, 75 (Suffolk: Boydell, 1984).

Stobart, Jon, 'Identity, Competition and Place Promotion in the Five Towns', *Urban History*, 30/2 (2003): 163–82.

Sweet, Rosemary, *Antiquaries. The Discovery of the Past in Eighteenth Century Britain* (London and New York: Hambledon, 2004).

Sykes, J.D., 'Agriculture and Science', in Mingay, Gordon Edmund, *The Victorian Countryside*, 2 volumes (London: Routledge and Kegan Paul, 1981).

Thackray, Arnold, 'Natural Knowledge in Cultural Context: The Manchester Model', *American Historical Review*, 79/3 (1974): 672–709.

Thompson, Francis Michael Longstreth, 'Pusey, Philip (1799–1855)', *Oxford Dictionary of National Biography* (Oxford: Oxford University Press, 2004).

Torrens, Hugh Simon, 'The Four Bath Philosophical Societies, 1779–1959', in Rolls, Roger, Guy, Jean, and Guy, John Richard (eds), *A Pox on the Provinces: Proceedings of the Twelfth Congress of the British Society for the History of Medicine* (Bath: Bath University Press, 1990).

Trainor, Richard, 'Peers on an Industrial Frontier: The Earls of Dartmouth and of Dudley in the Black Country, c.1810–1914', in Cannadine, David (ed.), *Patricians, Power and Politics in Nineteenth-Century Towns* (Leicester: Leicester University Press, 1982).

Tresidder, George A., 'Coronation Day Celebrations in English Towns, 1685–1821', *British Journal for Eighteenth Century Studies*, 15/1 (1992): 1–16.

Turner, Frank M., 'Public Science in Britain, 1880–1919', *Isis,* 71/4 (1980): 589–608.

Vyner, Blaise (ed.), *Building on the Past. Papers Celebrating 150 Years of the Royal Archaeological Institute* (London: Royal Archaeological Institute, 1994).

Walker, Andrew, 'The Lincolnshire Show, 1869–1913: Rural and Urban Encounters on the Showground', in K. Grieves et. al. (eds), *Rural and Urban Encounters in the Nineteenth and Twentieth Centuries: Regional Perspectives* (Conference of Regional and Local Historians, 2004).

Walsh, A.D., 'Notes on the First Visit of the British Association to Dundee and of the Early History of University College Dundee', in S.J. Jones (ed.), *Dundee and District* (Dundee: Dundee Local Executive Committee of the British Association, 1968).

Ward, Stephen V., *Selling Places: The Marketing and Promotion of Towns and Cities, 1850–2000* (London: E. and F.N. Spon, 1998).

Welch, Charles, 'Chance, Sir James Timmins, First Baronet (1814–1902)', Rev. A. McConnell, *Oxford Dictionary of National Biography* (Oxford: Oxford University Press, 2004).

Wetherall, David, 'The Growth of Archaeological Societies', in Brand, Vanessa (ed.), *The Study of the Past in the Victorian Age* (Oxford: Oxbow Books, 1998).

Williams, H.G., '"All Advance from Barbarism to Civilization is the Development of Social Science": Wales and the Promotion of Social Science', *Welsh History Review*, 20/4 (2001): 717–42.

Willliams, W.R., 'Vivian, Henry Hussey, First Baron Swansea (1821–1894)', Rev. Edmund Newell, *Oxford Dictionary of National Biography* (Oxford: Oxford University Press, 2006).

Withers, Charles W.J. and Finnegan, Diarmid A., 'Natural History Societies, Fieldwork and Local Knowledge in Nineteenth-century Scotland: Towards a Historical Geography of Civic Science', *Cultural Geographies*, 10 (2003): 334–53.

Withers, Charles W.J., Higgitt, Rebekah, and Finnegan, Diarmid A., 'Historical Geographies of Provincial Science: Themes in the Setting and Reception of the British Association for the Advancement of Science in Britain and Ireland, 1831–1939', *British Journal of the History of Science*, 41/3 (2008): 385–415.

Withers, Charles W.J., *Geography and Science in Britain, 1831–1939. A Study of the British Association for the Advancement of Science* (Manchester: Manchester University Press, 2010).

Wood, Alfred C., *A History of the University College of Nottingham, 1881–1948* (Oxford: Blackwell, 1953).

Wright, Gerry, 'Discussions of the Characteristics of Mechanics' Institutes in the Second Half of the Nineteenth Century: The Bradford Example', *Journal of Educational Administration and History*, 33/1 (2001): 1–16.

Wyllie, M.S., 'A Brief History of the Founding and Development of the Powysland Club', *Montgomeryshire Collections*, 86 (1998), pp. 123–43.

Young, G.M. (ed.), *Early Victorian England, 1830–1865. Volume II* (London: Oxford University Press, 3rd edition, 1963).

Unpublished Theses

Ebbatson, Linda, 'Conditions of Emergence and Existence of Archaeology in the Nineteenth Century: The Royal Archaeological Institute, 1843–1914' (unpublished Ph.D., University of Durham, 1999).

Goddard, Nicholas P.W., 'The Royal Agricultural Society of England and Agricultural Progress, 1838–1880' (unpublished Ph.D. thesis, University of Kent, 1981).

Kitteringham, G.S., 'Studies in the Popularisation of Science in England, 1800–1830' (unpublished Ph.D. thesis, University of Kent, 1981).

Works of Reference

Oxford English Dictionary, online version
Williams, John, *Digest of Welsh Historical Statistics*, vol. I (Cardiff: University of Wales Press, 1985).

Index

For Product Safety Concerns and Information please contact our EU representative GPSR@taylorandfrancis.com Taylor & Francis Verlag GmbH, Kaufingerstraße 24, 80331 München, Germany